Oklahoma Notes

Basic Sciences Review for Medical Licensure
Developed at
The University of Oklahoma College of Medicine

Suitable Reviews for:
United States Medical Licensing Examination
(USMLE), Step 1

Oklahoma Notes

Neuropathology and Basic Neuroscience

Roger A. Brumback
Richard W. Leech

Springer-Verlag
New York Berlin Heidelberg London Paris
Tokyo Hong Kong Barcelona Budapest

Roger A. Brumback, M.D.
Department of Pathology
University of Oklahoma
Biomedical Sciences Building,
  Room 451
940 Stanton L. Young Boulevard
Oklahoma City, OK 73104
USA

Richard W. Leech, M.D.
Department of Pathology
University of Oklahoma
Biomedical Sciences Building,
  Room 451
940 Stanton L. Young Boulevard
Oklahoma City, OK 73104
USA

Library of Congress Cataloging-in-Publication Data
Brumback, Roger A.
    Neuropathology and basic neuroscience / Roger A. Brumback, Richard
W. Leech.
      p.   cm. — (Oklahoma notes)
    Includes bibliographical references.
    ISBN 0-387-94389-7
    1. Nervous system—Diseases—Outlines, syllabi, etc.
    2. Neurosciences—Outlines, syllabi, etc.   I. Leech, Richard W.
    II. Title.   III. Series.
      [DNLM: 1. Nervous System Diseases—outlines.   2. Nervous System
    Diseases—programmed instruction.   WL 18 B893na 1995]
    RC357.B79   1995
    616.8—dc20
    DNLM/DLC
    for Library of Congress   94-42193

Printed on acid-free paper.

Production managed by Jim Harbison; manufacturing supervised by Jacqui Ashri.
Camera-ready copy prepared by the author.

9 8 7 6 5 4 3 2 1

ISBN 0-387-94389-7 Springer-Verlag New York Berlin Heidelberg

Preface to the
Oklahoma Notes

In 1973, the University of Oklahoma College of Medicine instituted a requirement for passage of the Part 1 National Boards for promotion to the third year. To assist students in preparation for this examination, a two-week review of the basic sciences was added to the curriculum in 1975. Ten review texts were written by the faculty: four in anatomical sciences and one each in the other six basic sciences. Self-instructional quizzes were also developed by each discipline and administered during the review period.

The first year the course was instituted the Total Score performance on National Boards Part I increased 60 points, with the relative standing of the school changing from 56th to 9th in the nation. The performance of the class since then has remained near the national candidate mean. This improvement in our own students' performance has been documented (Hyde et al: Performance on NBME Part I examination in relation to policies regarding use of test. J. Med. Educ. 60: 439–443, 1985).

A questionnaire was administered to one of the classes after they had completed the Boards; 82% rated the review books as the most beneficial part of the course. These texts were subsequently rewritten and made available for use by all students of medicine who were preparing for comprehensive examinations in the Basic Medical Sciences. Since their introduction in 1987, over 300,000 copies have been sold. Obviously these texts have proven to be of value. The main reason is that they present a *concise overview* of each discipline, emphasizing the content and concepts most appropriate to the task at hand, i.e., passage of a comprehensive examination over the Basic Medical Sciences.

The recent changes in the licensure examination that have been made to create a Step 1/Step 2/Step 3 process have necessitiated a complete revision of the Oklahoma Notes. This task was begun in the summer of 1991 and has been on-going over the past 3 years. The book you are now holding is a product of that revision. Besides bringing each book up to date, the authors have made every effort to make the tests and review questions conform to the new format of the National Board of Medical Examiners. Thus we have added numerous clinical vignettes and extended match questions. A major revision in the review of the Anatomical Sciences has also been introduced. We have distilled the previous editions' content to the details the authors believe to be of greatest importance and have combined the four texts into a single volume. In addition a book over neurosciences has been added to reflect the emphasis this interdisciplinary field is now receiving.

I hope you will find these review books valuable in your preparation for the licensure exams. Good Luck!

Richard M. Hyde, Ph.D.
Executive Editor

Preface

The United States Congress designated the 1990's as the "Decade of the Brain" in recognition of the importance of neuroscience to the health and well-being of Americans. It has been suggested that as many as 20% of all patients seeking medical treatment have neurologic problems, either as the presenting complaint or as an associated condition complicating the primary illness. To this end, it is important that physicians understand basic neuroscience principles and nervous system diseases. Of course, this text is not encyclopedic but instead is an outline of the knowledge required of all medical students. Interested students can consult numerous texts that provide comprehensive coverage of the field, including *Greenfield's Neuropathology* and the exhaustive 60+ volume *Handbook of Clinical Neurology*. The information selected for inclusion in this volume of the *Oklahoma Notes* series remains true to the goal of the whole series—incorporating only that material vital in both the general clinical practice of medicine and to answer questions on the all-important United States Medical Licensing Examination.

Roger A. Brumback
Richard W. Leech

Acknowledgments

This text would not have been possible without a great deal of help and support from a number of individuals. We want to thank all those who assisted in our education in neuroscience and neuropathology including: William M. Landau and Philip R. Dodge of the Washington University School of Medicine, Lowell W. Lapham of the University of Rochester Medical Center, and Ellsworth C. Alvord, Jr. of the University of Washington School of Medicine. These great teachers demonstrated the importance for good patient care of a thorough understanding of the pathophysiologic mechanisms of disease. Currently, Fred G. Silva, Chairman of the Department of Pathology of the University of Oklahoma College of Medicine has provided us with the supportive environment necessary to pursue our teaching and writing.

In the preparation of this volume, Elizabeth Claire Maletz supplied the many drawings and Mary Helen Brumback assisted with the typing and proofreading.

Contents

CHAPTER 1: NERVOUS SYSTEM STRUCTURE AND FUNCTION

I. Components

    A. Central nervous system — brain, spinal cord, olfactory tract and bulb, optic nerve and retina

        1. **Gray matter**

            a. Regions that contain large numbers of nerve cell bodies; so designated because of gross color

            b. **Neuropil** — ramifications of dendrites and terminals of axons forming network around neuronal cell bodies

        2. **White matter** — regions composed mainly of myelinated fibers; so designated because of gross color

    B. Peripheral nervous system — spinal and cranial nerve roots, sensory and autonomic ganglia, peripheral nerves, peripheral sensory receptors

II. Central nervous system subdivisions

    A. **Spinal cord**

        1. Elongated cylindrical structure within vertebral canal extending from foramen magnum approximately to level of second lumbar vertebra

        2. Connecting structure between brain and periphery; conveys sensory information from and sends motor impulses to peripheral regions

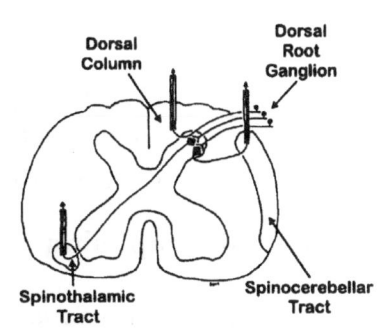

Major spinal cord sensory tracts.

3. Mediates many reflexes connecting sensory input with motor output without involving more rostral (higher) central nervous system levels (brain)

B. **Brain stem**

1. Expanded rostral extension of spinal cord above foramen magnum in posterior cranial fossa

2. Mediates same functions for face that are carried out by spinal cord for rest of body

3. Contains auditory and eye movement systems

4. **Medulla** — most caudal portion of brain stem, connecting with spinal cord through foramen magnum

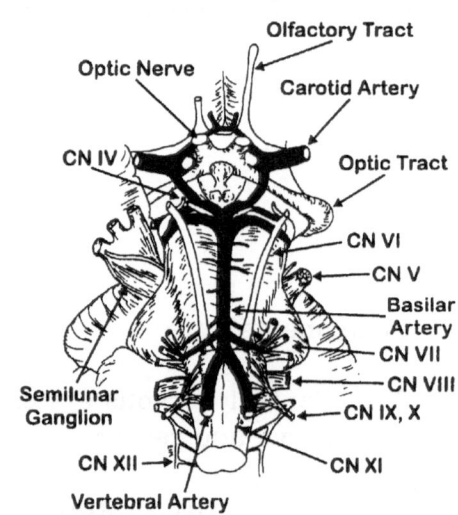

Brain stem with cranial nerves and arteries.

5. **Pons** — middle portion of brain stem, rostral to medulla and anterior to cerebellum

6. **Midbrain** (mesencephalon) — rostral continuation of pons, lying in tentorial incisura (notch) and connecting posterior cranial fossa structures (pons and cerebellum) with supratentorial structures (diencephalon)

C. **Cerebellum**

1. Occupies most of posterior cranial fossa

2. Coordinates volitional motor activity by integrating sensory information conveyed through pons and medulla with motor instructions from basal ganglia and cerebral cortex

D. **Diencephalon**

1. Most caudal supratentorial structure, consisting of paired (left and right) sets of nuclei: thalamus, hypothalamus, epithalamus, subthalamus, lateral and medial geniculates

2.   **Thalamus** — relays **sensory** information and other impulses from brain stem and other structures to cerebral cortex and interacts with cerebral cortex

3.   **Hypothalamus** — controls **endocrine function** and other vegetative functions necessary for life (metabolism of food, salt and water balance, reproduction, temperature control, responses to threat and stress)

4.   **Subthalamus** — interacts with basal ganglia in control of movement

5.   **Epithalamus** — control of body rhythms through interaction with pineal gland

6.   **Lateral geniculate** — relays visual input received from optic tract to visual cortex (occipital lobe)

7.   **Medial geniculate** — relays auditory input received from brain stem to auditory cortex (temporal lobe)

E.   **Basal ganglia**

1.   Gray matter structures deep within each cerebral hemisphere; important in controlling motor function

2.   **Striatum** — **caudate** nucleus and **putamen** which are continuous anteriorly

3.   **Globus pallidus**

F.   Corticoid and allocortex

1.   Primitive cerebral structures with rudimentary layering (stratification); phylogenetic precursors of six-layered cerebral isocortex (cerebral cortex that comprises majority of cerebral hemisphere)

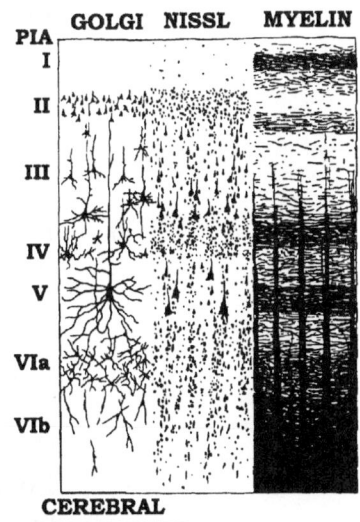

Appearance of six-layered cerebral isocortex in Golgi, Nissl, and myelin stained sections.

2.   Includes **amygdala**, substantia innominata (**nucleus basalis of Meynert**), piriform (olfactory) cortex, **hippocampus**

3. Controls memory functions and **mediates interactions** between rest of **cerebral cortex** and **hypothalamic control** of endocrine functions and essential activities necessary to life

G. Cerebral cortex

1. Largest part of human brain; involved in language, abstract thinking, perception, and control of movement and adaptive responses to outside world

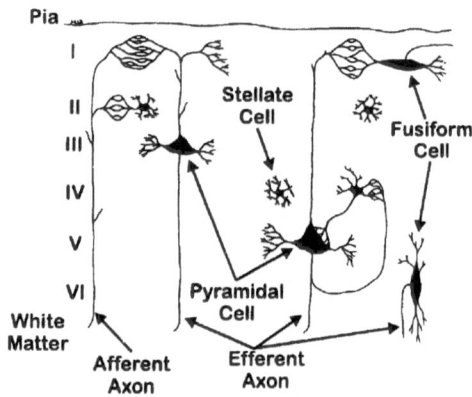

Arrangement of major neuronal types in cerebral isocortex.

2. **Cerebral isocortex (neocortex)**

    a. **Six-layered** cerebral gray matter

        (1) Layer I — **molecular layer** containing terminal dendrites and axons of cerebral cortical neurons; occasional horizontally-oriented fusiform neurons provide some interconnection

        (2) Layer II — **outer granular layer** containing many small granular (stellate) interneurons; receives input from other cerebral cortical regions

        (3) Layer III — **outer pyramidal cell layer** contains small to medium size pyramidal neurons which project to other cerebral cortical regions

        (4) Layer IV — **inner granular layer** containing many small granular (stellate) interneurons; receives major input from thalamus and other subcortical nuclei; contains **band of horizontally-oriented myelinated fibers** which is most prominent in primary visual cortex (**line of Gennari**)

        (5) Layer V — **inner pyramidal cell layer** contains **large pyramidal neurons** (including **Betz cells** in primary motor cortex); axons of these pyramidal cells project to subcortical sites (including brain stem, cerebellum, and spinal cord)

(6) Layer VI — **multiform layer** containing fusiform neurons that project to thalamus

b. Cytoarchitectonic areas — division of cerebral cortex into different areas based upon neuronal density; most popular such system developed by Brodmann

(1) Areas 3, 1, 2 — **postcentral gyrus** containing primary somatosensory cortex

(2) Area 4 — **precentral gyrus** containing primary motor cortex which gives rise to corticospinal tract (**Betz cells** present in this area)

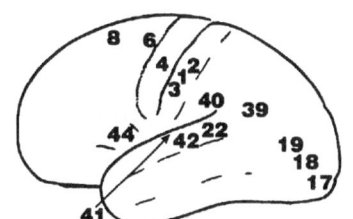

Representative cytoarchitectonic areas according to Brodmann numbering system for specific areas of cerebral isocortex.

(3) Area 6 — premotor area anterior to area 4; projects fibers to area 4 and also contributes to corticospinal tract

(4) Area 8 — **frontal eye fields**; responsible for voluntary conjugate eye deviation to opposite side

(5) Area 44 — **Broca's area**, immediately in front of motor areas for face (lower ends of areas 4 and 6); lesions of this area are associated with expressive aphasia

(6) Area 41 — **Heschl's gyrus** (transverse temporal gyrus) lying horizontally in Sylvian fissure contains primary auditory cortex

(7) Areas 42, 22 — auditory association cortex associated with receptive language functions

(8) Area 40 — supramarginal gyrus; lesions of this area are associated with agnosia and aphasia syndromes

(9) Area 39 — angular gyrus; lesions of this area are associated with complex behavioral disturbances (lesions in

6

left hemisphere are associated with Gerstmann syndrome, consisting of acalculia, agraphia, finger agnosia, and right-left disorientation)

(10) Area 17 — **primary visual cortex (striate cortex; calcarine cortex)**

(11) Areas 18, 19 — visual association cortex

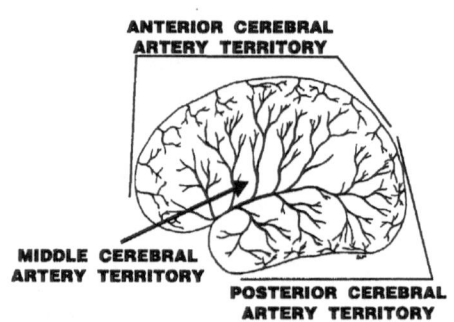

Distribution of cerebral arteries over lateral convexity.

III. Blood supply

    A. Arterial supply of brain

        1. Four arteries supply blood to intracranial contents: paired internal carotid arteries and paired vertebral arteries

        2. **Circle of Willis — anastomosis** between vascular supplies from bilateral internal carotid arteries and bilateral vertebral arteries

        3. **Carotid circulation** — supplies most of supratentorial compartment, through anterior cerebral artery and middle cerebral artery

        4. **Vertebrobasilar circulation** — vertebral arteries join just rostral to foramen magnum to form basilar artery which supplies posterior fossa structures; rostral basilar artery bifurcates at tentorial incisura to form right and left posterior cerebral arteries which enter supratentorial compartment to supply most of posterior cerebrum

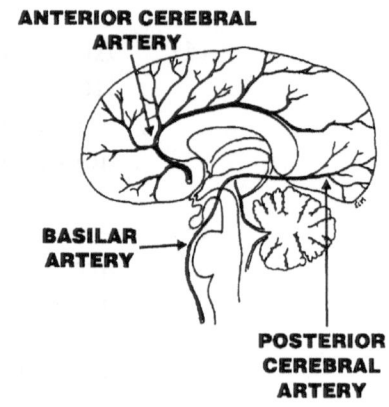

Distribution of cerebral arteries on medial surface of hemisphere.

    B. Venous drainage of brain

        1. **Dural venous sinuses** — endothelial-lined venous channels within folds of dura

2. Brain venous channels do not have valves: blood flows in either direction depending on momentary pressure changes

3. **Superior sagittal sinus** — superficial **cerebral cortical veins** empty into superior sagittal sinus which then empties into confluence of sinuses

4. Deep venous system

    a. Each cerebral hemisphere is drained by internal cerebral vein which is confluence of thalamostriate vein, septal vein, and choroidal vein

    b. Right and left internal cerebral veins join to form **great vein of Galen** just posterior and inferior to splenium of corpus callosum

    c. **Great vein of Galen** empties into anterior end of **straight sinus**, which is contained in dural folds of tentorium

    d. Straight sinus and superior sagittal sinus merge at confluence of sinuses

    e. Confluence of sinuses drains into right and left transverse sinuses, which become right and left sigmoid sinuses that then drain into **internal jugular veins**

C. **Middle meningeal artery**

1. Periosteal branch of **external carotid artery** that enters cranial cavity through foramen spinosum and ramifying in grooves along inside of skull vault

2. Provides blood supply to dura and skull bones, but ordinarily supplies no central nervous system tissue

D. **Spinal arteries**

1. At level of foramen magnum, each vertebral artery gives off branches to supply spinal cord

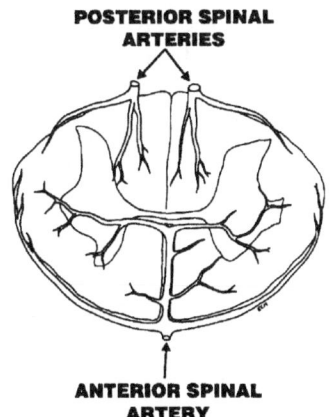

Distribution of spinal arteries.

8

      a.   Posterior spinal arteries run caudally near each dorsal root forming plexiform network that supplies dorsal columns, substantia gelatinosa of dorsal horns, and dorsal root entry zone

      b.   Anterior branches join in midline to form single **anterior spinal artery** that supplies anterior and lateral portions of spinal cord

   2.   Radicular arteries

      a.   Radicular branches from descending aorta pass through vertebral foramina to join with anterior and posterior spinal arteries, adding to blood supply for lower portions of spinal cord

      b.   **Artery of Adamkiewicz** — largest radicular branch (at lower thoracic level)

IV.   **Neuron** (nerve cell) — specialized, long-lived (life-long and does not replicate), multibranched cell with electrically-excitable membrane for reception, integration, and communication of information

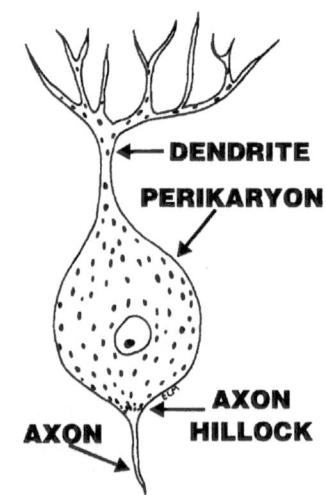

   A.   Shape varies from stellate to pyramidal (triangular or flask-shaped); size varies from about 125 μm diameter (spinal cord anterior horn motor neurons and Betz cells of precentral gyrus) to about 4 μm in diameter (cerebellar granule cells)

   B.   **Perikaryon** (nerve cell body; soma; sometimes referred to simply as "neuron")

Cerebellar Purkinje cell.

      1.   **Nucleus — contains prominent nucleolus**, which is useful feature to distinguish nerve cells from other nervous system cells in histologic sections

      2.   **Nissl substance**

         a.   **Basophilic** (blue-staining with hematoxylin and eosin stain) **granules** in nerve cell perikaryon

         b.   Fills cytoplasm of nerve cell perikaryon and **extends into dendrites**, but is **excluded from axon hillock and axon**

c.   Composed of concentrated collections of **ribosomes and endoplasmic reticulum** for protein synthesis

3.   Cytoplasmic pigment

    a.   **Lipofuscin** — brown pigment found in lysosomes; amount increases with advancing age

    b.   **Neuromelanin** — melanin in lysosomes (distinguished from melanocyte melanin which is contained in specialized structures called melanosomes); found in aminergic neurons of **substantia nigra** (midbrain), **locus ceruleus** (pons), and dorsal motor nucleus of vagus (medulla)

C. **Axon**

1.   Slender, cylindrical, uniform diameter process arising from perikaryon at axon hillock; only one axon from each neuron

2.   Ensheathed by supporting cells (Schwann cells in peripheral nervous system or oligodendrocytes in central nervous system), which either make myelin (myelinated axons) or do not make myelin (unmyelinated axons); myelinated axons are larger (up to 20 μm) in diameter than unmyelinated axons (less than 2 μm)

3.   Axoplasmic transport — axon contains numerous neurotubules (axonal microtubules) and neurofilaments, which are involved in movement of material between perikaryon and terminal portions of axon

    a.   **Anterograde slow transport** — movement of **cell organelles** such as lysosomes, mitochondria, and vesicles toward axon terminals at rate of about **1 to 3 mm/day**

    b.   **Anterograde fast transport** — movement of **peptides, proteins,** and **metabolites** (including neurotransmitters or precursors) toward axon terminals at rate of about **400 mm/day**

    c.   **Retrograde transport** — movement of materials from axon terminals toward cell body at varying rate of up to about 400 mm/day

4.   Conveys (propagates) electrical signal (depolarization) to distant points in order to effect receptors on target structure (other neuronal processes or peripheral organs such as muscle)

    a.   **Synapses**

        (1)   Structure composed of **terminal portion of axon** (axon terminal), **membrane of target structure** (receptor site), and intervening space (synaptic cleft)

        (2)   Synapses can be formed between various structures:

            (a)   Axodendritic — axon to dendrite

            (b)   Axosomatic — axon to perikaryon

            (c)   Axoaxonic — axon to axon

            (d)   Neuromuscular — axon to muscle

        (3)   **Synaptic vesicles** — vesicles in axon terminal which store chemical neurotransmitters that can be released into synaptic cleft

        (4)   Axon effects target receptors by releasing **chemical neurotransmitter** into synaptic cleft where it diffuses toward cell membrane of target structure in order to interact with specialized receptors on surface of target membrane

    b.   Gap junctions — axon terminal membrane that is so closely approximated to target cell membranes that direct electrical transmission of depolarization potential is possible

    c.   **Neurosecretory neurons** — axon terminals synapse in close proximity to blood vessels where neuropeptides (hormones) are released to enter the blood circulation; such cells are particularly prominent in hypothalamus

D. **Dendrite**

    1.   Short cell process (often multiple) extending from neuronal perikaryon

    2.   Specialized to receive signals from axon terminals (synapses)

    3.   **Dendritic spines** — short fine branches of dendrites that serve as points of attachment of synapses

4.   Cytoplasm of dendrites similar to that of perikaryon (including presence of Nissl substance)

E.   Some specific neuronal types

1.   **Anterior horn cell** ("lower motor neuron") — **motor neuron** in spinal cord anterior horn that sends axon through ventral spinal root and peripheral nerve to synapse on skeletal muscle fibers

2.   **Betz cell** ("upper motor neuron") — large pyramidal neuron in cerebral cortical **precentral gyrus**; axon travels in corticospinal tract to synapse on anterior horn cells

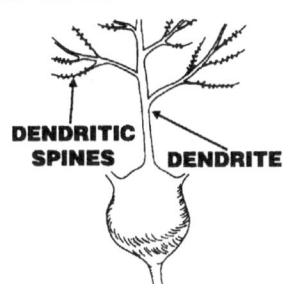

Spines on neuronal dendrites.

3.   **Purkinje cell** — large **cerebellar** neuron located in single layer between cerebellar granule cells and molecular layer; axon synapses in cerebellar dentate nucleus

F.   **Neuronal physiology**

1.   Ion channel structure

a.   **Voltage-gated ion channels** (for Na^+, K^+, Cl^-, Ca^{2+}) are present in varying amounts along whole membrane of neurons, while **ligand-gated** ion channels are present at synapses

b.   **Selectivity filter** — channel has narrowed point near exterior opening of channel that "selects" ion permitted to pass through that particular type of channel

c.   **Voltage sensor** — charged proteins that "sense" voltage changes and alter conformation such that gate near interior opening of channel swings open allowing ions to pass through channel

d.   **Pore** — channel through lipid bilayer of membrane that allows ions to travel between interior and exterior of cell

2.   **Ion channel function**

a.   Nerve cell membrane is **selectively permeable** to ions; major intracellular ion is potassium and major extracellular ion is sodium resulting in intracellular charge (resting potential) of about –70 mV compared to extracellular environment

(1) Membrane ion pumps maintain charge by maintaining ion gradients; major ion pump is ATPase-dependent sodium-potassium pump that transports sodium ions out of cell in exchange for pumping potassium ions into cell

(2) **Hyperpolarization** — change in transmembrane potential toward a **higher negative value**

(3) **Depolarization** — change in transmembrane potential toward **less negative or positive value**

b. **Action potential**

(1) **Depolarization** — reduction of local transmembrane potential below threshold level (about –50 mV) results in precipitous change in local membrane permeability with opening of voltage-gated sodium ion channels allowing explosive passage of sodium ions such that transmembrane potential changes toward +50 mV

(2) **Repolarization** — open sodium channels are quickly closed (inactivated), while voltage-gated potassium ion channels open allowing efflux of potassium ions to balance earlier sodium ion influx, thereby repolarizing membrane toward resting potential

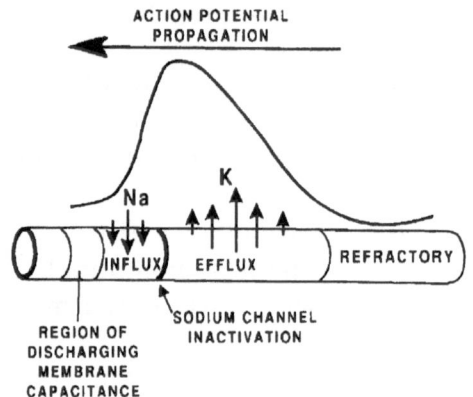

Action potential propagation along excitable membrane.

(3) Since potassium ion channels are only slowly closed (inactivated), repolarization overshoots resting membrane potential resulting in hyperpolarization and transient **refractory period** during which membrane is refractory to further depolarization (regardless of size of stimulating current)

(4) Current flow resulting from depolarization event triggers depolarization of adjacent areas of membrane, ultimately resulting in action potential that propagates over whole length of membrane

c. **Neurotransmitter release**

(1) **Quanta** — quantity of neurotransmitter stored in synaptic vesicles in presynaptic axon terminals and released as single unit into synaptic cleft

(2) Depolarization current reaching axon terminal causes **transient opening of voltage-gated calcium channels** which allows entry of calcium ions into axon terminal

(3) **Release of neurotransmitter occurs by exocytosis** — free calcium ions in axon terminal cause synaptic vesicles to fuse with plasma membrane of axon terminal (presynaptic membrane) releasing neurotransmitter into synaptic cleft

d. Pharmacologic interaction with sodium channels

(1) **Tetrodotoxin** (puffer fish poison) or saxitoxin (paralytic shellfish poison) — bind to exterior end of channel at selectivity filter, blocking sodium ion passage

(2) **Local anesthetics** (such as lidocaine or procaine) — bind in channel pore, blocking sodium ion passage

e. Postsynaptic potentials

(1) Neurotransmitter released by axon terminal (presynaptic membrane) into synaptic cleft interacts with receptors on target cell (postsynaptic membrane) to alter ionic channels and produce localized change in membrane potential

(2) Two types — **excitatory postsynaptic potential (EPSP) or inhibitory postsynaptic potential (IPSP)**

(a) **IPSP**

i) **Decreases excitability** of postsynaptic membrane by hyperpolarizing membrane

        ii)   Opening of ion channels that allow movement of chloride ions into cell (in some situations also to permit potassium ion movement out of cell) makes membrane potential more negative

  (b) **EPSP**

        i)   **Increases excitability** of postsynaptic membrane by depolarizing membrane

        ii)   Opening of ion channels that allow movement of sodium ions into cell makes membrane potential more positive

(3)   Each postsynaptic potential produces localized, transient, nonpropagated change in membrane potential

  (a)   Localized postsynaptic potentials (either IPSP or EPSP) are transient, lasting only about 10-15 msec and changing membrane potential by about 0.1 mV

  (b)   Multiple localized postsynaptic potentials occurring nearly simultaneously (due to several adjacent presynaptic axon terminals concurrently releasing neurotransmitter) are summated

  (c)   Propagated potential (**action potential**) results when sufficient number (generally more than 100) of **simultaneous excitatory potentials** (EPSP) can overcome effect of any simultaneous inhibitory potentials (IPSP) and summate to produce change in postsynaptic membrane potential that exceeds threshold for generation of action potential

G. Neurotransmitters

1. Acetylcholine

  a. **Excitatory neurotransmitter** synthesized by **choline acetyltransferase** from acetyl CoA and choline

Acetylcholine.

b.   Major pathways

 (1) **Motor neurons** with axon terminals at neuromuscular junction

 (2) **Autonomic postganglionic parasympathetic neurons** with axon terminals on effector organs

 (3) **Basal forebrain cholinergic system** — basal forebrain neurons (substantia innominata, **nucleus basalis of Meynert**, septal nuclei, nucleus of diagonal band of Broca) that have axonal projections to hippocampus and rest of cerebral cortex

 (4) **Striatal interneurons** that have axonal terminals within striatum

 (5) Pontomesencephalic cholinergic system — pontomesencephalic neurons that have axonal projections to diencephalon, pontine reticular formation, and cerebellum

c.   Catabolism

 (1) Broken down in synaptic cleft by **acetylcholinesterase**, which hydrolyzes acetylcholine into acetate and choline

 (2) Hydrolyzed by **nonspecific cholinesterase** (butyrylcholinesterase; "pseudocholinesterase") found in non-neural tissues and blood

 (3) Choline reuptake — choline in synaptic cleft is transported through presynaptic membrane into presynaptic terminal to be resynthesized into acetylcholine

d.   **Acetylcholine receptors**

 (1) **Muscarinic receptors** — five subtypes (M_1, M_2, M_3, M_4, M_5); linked to guanine nucleotide-binding regulatory proteins (G-proteins); depending on cell type, produces either depolarization or hyperpolarization of membrane

16

(2) **Nicotinic receptors**

- (a) Four subtypes — CNS presynaptic, CNS postsynaptic, autonomic ganglionic, skeletal muscle

- (b) In skeletal muscle — **opens sodium channels** (ligand-gated cation channel) resulting in membrane depolarization

e. Pharmacology

(1) **Anticholinesterases**

- (a) **Inhibition of cholinesterase** — results in increased concentration of acetylcholine in synaptic cleft

- (b) Inhibit enzyme by acylating esteratic site on enzyme

  - i) **Reversible inhibitor (physostigmine,** pyridostigmine, edrophonium) — competitive with acetylcholine

  - ii) **Irreversible inhibitor (organophosphorus insecticides** and nerve gases) — irreversibly phosphorylates esteratic site; **hydroxylamine compounds** (such as **2-PAM**) can displace inhibitor (thereby "reversing" irreversible inhibitor)

(2) **Muscarinic blocker — atropine** blocks **peripheral autonomic muscarinic receptors,** but cannot cross blood-brain barrier; scopolamine can cross blood-brain barrier and therefore can block both peripheral and central receptors

(3) **Nicotinic blocker**

- (a) **Curare** and hexamethonium block peripheral nicotinic receptors, resulting in neuromuscular paralysis (muscle relaxants)

- (b) **Succinylcholine** — nicotinic receptor agonist that has greater affinity for receptor than does acetylcholine: after producing initial receptor depolarization ("depolarizing blocker"), succinylcholine remains attached to receptor preventing normal receptor recovery and blocking access by acetylcholine; thus, neuromuscular paralysis ensues

2. **Dopamine**

Dopamine.

a. **Monoamine neurotransmitter** synthesized from **L-dopa** (L-dihydroxyphenylalanine) by pyridoxine-dependent enzyme **dopa decarboxylase**; L-dopa is synthesized from tyrosine by enzyme tyrosine hydroxylase

b. Major pathways

(1) Short **interneurons in retina and olfactory bulb**

(2) **Tuberoinfundibular dopaminergic system** — neurons in hypothalamus that have axon terminals in median eminence, releasing dopamine into blood vessels which subsequently ramify in anterior pituitary; dopamine **inhibits release of prolactin**

(3) **Nigrostriatal dopaminergic system** — neurons in substantia nigra that have axonal projections to striatum

(4) Mesocortical-mesolimbic dopaminergic system — neurons in substantia nigra and midbrain ventral tegmental area that have axonal projections to limbic structures (limbic cortex, nucleus accumbens septi, amygdala, piriform cortex)

c. Catabolism

(1) **Reuptake** — after release into synaptic cleft, reuptake pump on presynaptic membrane (high-affinity uptake sites) transports dopamine back

HVA (homovanillic acid).

into axon terminal to be repackaged in synaptic vesicles or metabolized further

(2) Free dopamine metabolized by intraneuronal enzyme **monoamine oxidase** and extraneuronal enzyme **catechol-*O*-methyltransferase** to **homovanillic acid (HVA)**

d. **Dopaminergic receptors**

(1) **D_1 postsynaptic receptor family** (subtypes D_1, D_5) — stimulates adenylyl cyclase activity and phosphoinositide turnover

(2) **D_2 postsynaptic receptor family** (subtypes D_2, D_3, D_4) — inhibits adenylyl cyclase activity, inhibits calcium ion entry through calcium channels, enhances potassium ion passage through potassium channels

(3) **Dopamine autoreceptors** — receptors on perikaryon, dendrites, and axons of dopamine-containing neurons; action is to inhibit dopamine synthesis and release

e. Pharmacology

(1) Dopamine depletion — **reserpine** or **tetrabenazine interferes with storage of dopamine in vesicles**

(2) Dopamine agonist — **bromocriptine** or **apomorphine** directly stimulate (agonist for) dopamine receptors

(3) Dopamine antagonist — **antipsychotic drugs** (major tranquilizers) such as **chlorpromazine** or **haloperidol** block dopamine receptors, preventing receptor interaction with dopamine in synaptic cleft

(4) Reuptake blocker — **amphetamine** or **cocaine** blocks reuptake mechanism for dopamine, increasing dopamine concentration in synaptic cleft

(5) Monoamine oxidase inhibition — **iproniazid, pargyline,** or **deprenyl** inhibit monoamine oxidase, increasing available dopamine

3.   **Norepinephrine**

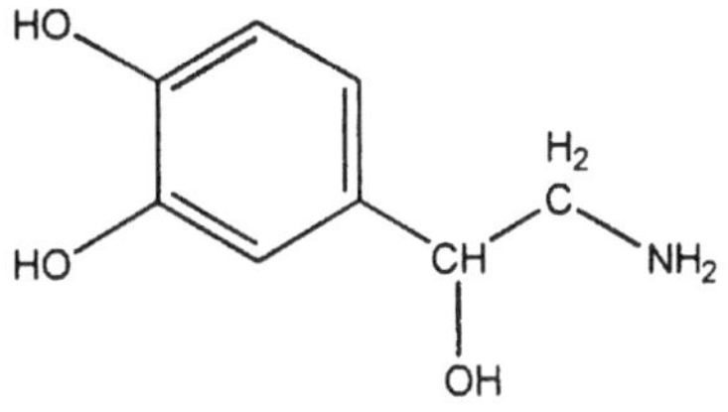

a.   **Monoamine neurotransmitter** synthesized from **dopamine** by enzyme **dopamine-ß-hydroxylase**

Norepinephrine.

b.   Major pathways

(1)   Autonomic postganglionic sympathetic neurons with axon terminals on effector organs

(2)   Released by adrenal medullary cells (embryologically, these cells are similar to postganglionic sympathetic neurons)

(3)   Locus ceruleus-lateral tegmental norepinephrine system — neurons in pontine locus ceruleus and adjacent pontine tegmentum have axon projections to whole cerebral cortex, diencephalon, cerebellum, and spinal cord

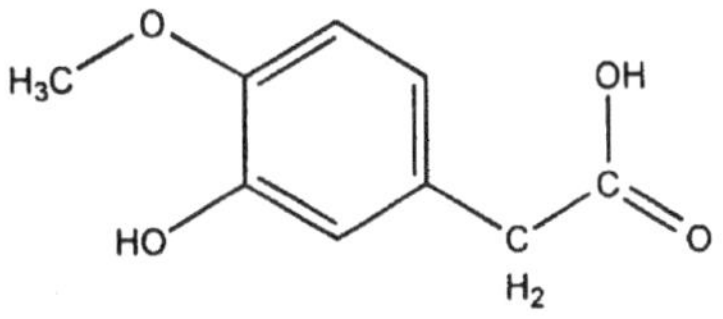

VMA (vanillylmandelic acid).

c.   Catabolism

(1)   After release into synaptic cleft, reuptake pump on presynaptic membrane (high-affinity uptake sites) transports norepinephrine back into axon terminal to be repackaged in synaptic vesicles or metabolized further

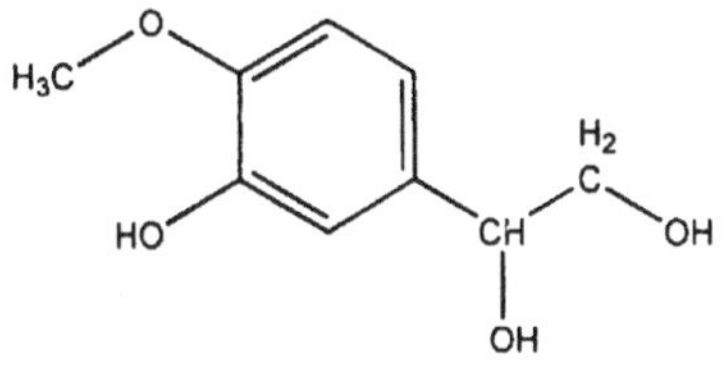

MHPG (3-methoxy-4-hydroxyphenylglycol).

(2)   Free norepinephrine metabolized by intraneuronal enzyme **monoamine oxidase** and extraneuronal enzyme **catechol-*O*-methyltransferase** to **vanillylmandelic acid (VMA)** or **3-methoxy-4-hydroxyphenylglycol (MHPG)**

d. **Adrenergic receptors** — action coupled to G-proteins; in brain, stimulation of adrenergic receptors inhibits spontaneous neuronal discharges

(1) α_1 receptor (subtypes α_{1A}, α_{1B}, α_{1C}) — postsynaptic receptor that activates phospholipase C and increases intracellular calcium ions; in peripheral tissues contracts smooth muscle and is located in blood vessels and heart; also, found in brain

(2) α_2 receptor (subtypes α_{2A}, α_{2B}, α_{2C}) — inhibits adenylyl cyclase activity; in peripheral tissues, mainly localized to presynaptic membrane where receptor stimulation serves to block release of norepinephrine

(3) β_1 receptor — found in high density in cerebral cortex and heart; activates adenylyl cyclase; peripheral stimulation results in fatty acid mobilization from adipose tissue and increased cardiac function

(4) β_2 receptor — found in high density in lung and cerebellum; activates adenylyl cyclase; peripheral stimulation results in bronchodilation, glycogenolysis, and inhibition of uterine contraction

(5) β_3 receptor — found in brown adipose tissue; peripheral stimulation results in lipolysis

(6) Norepinephrine autoreceptors — receptors on perikaryon, dendrites, and axons of norepinephrine-containing neurons; action is to inhibit norepinephrine synthesis and release

e. Pharmacology

(1) Norepinephrine depletion — **reserpine** or **tetrabenazine interferes with storage of norepinephrine in vesicles**

(2) **Norepinephrine agonist** — **clonidine** stimulates autoreceptors, thereby blocking release of norepinephrine

(3) **Norepinephrine antagonist** — phenoxybenzamine and phentolamine block α-receptors, preventing receptor interaction with norepinephrine in synaptic cleft

(4) **Reuptake blocker — tricyclic antidepressants** (such as desipramine) block reuptake mechanism for norepinephrine, increasing norepinephrine concentration in synaptic cleft

(5) Monoamine oxidase inhibition — **iproniazid** or **pargyline** inhibit monoamine oxidase, increasing available norepinephrine

4. **Serotonin** (5-hydroxytryptamine; 5-HT)

a. **Monoamine neurotransmitter** synthesized from **5-hydroxytryptophan** by enzyme **aromatic amino acid decarboxylase**; 5-hydroxytryptophan is synthesized from tryptophan by **tryptophan hydroxylase**

Serotonin.

b. Major pathways

(1) Found in **platelets**, mast cells, and throughout gastrointestinal tract

(2) Found in high concentration in pineal cells, where it serves as precursor for melatonin synthesis

(3) **Brain stem raphe nuclei serotonergic system** — neurons in dorsal and median raphe nuclei (in rostral brain stem) have axon projections to whole cerebral cortex, basal ganglia, diencephalon, cerebellum, and spinal cord

c. Catabolism

5-Hydroxyindolacetic acid.

(1) After release into synaptic cleft, reuptake pump on presynaptic membrane (high-affinity uptake sites) transports serotonin back into axon terminal to be repackaged in synaptic vesicles or metabolized further

(2)   Free serotonin metabolized by intraneuronal enzyme **monoamine oxidase** to **5-hydroxyindoleacetic acid (5-HIAA)**

d.   **Serotonergic (5-HT) receptors**

(1) 5-HT_{1A} — found in highest density in raphe nuclei and hippocampus; action coupled to G-proteins, with resultant opening of potassium ion channels and hyperpolarization of postsynaptic membrane

(2) 5-HT_{1B} — action coupled to G-proteins with inhibition of adenylyl cyclase activity

(3) 5-HT_{1C} — action coupled to G-proteins with stimulation of phosphoinositide hydrolysis

(4) 5-HT_{1D} — action coupled to G-proteins with inhibition of adenylyl cyclase activity

(5) 5-HT_{2} — found in high concentration in cerebral cortex; action coupled to G-proteins with stimulation of phosphoinositide hydrolysis resulting in inhibition of potassium ion channels with consequent postsynaptic membrane depolarization

(6) 5-HT_{3} — found in high concentration in medullary **area postrema** (stimulation results in vomiting), and peripheral nerve ganglia and spinal cord substantia gelatinosa (mediating pain mechanisms); opens gated ion channel allowing influx of sodium ions (and other cations) with resultant depolarization of postsynaptic membrane

(7) 5-HT_{4} — localized in midbrain colliculi and hippocampus; activates adenylyl cyclase with consequent inhibition of potassium ion channels and depolarization of postsynaptic membrane

e.   Pharmacology

(1)   Serotonin depletion — **reserpine** or **tetrabenazine interferes with storage of serotonin in vesicles**

(2)   Serotonin agonist — **lysergic acid diethylamide (LSD)** stimulates serotonin receptors

(3)   **Reuptake blocker — tricyclic antidepressants** (such as imipramine or amitriptyline) block reuptake mechanism for serotonin, increasing serotonin concentration in synaptic cleft

(4)   Monoamine oxidase inhibition — **iproniazid** or **pargyline** inhibit monoamine oxidase, increasing available serotonin

5.   **γ-Aminobutyric acid (GABA)**

a.   **Inhibitory neurotransmitter** synthesized in GABA shunt pathway

GABA.

(1)   **Glucose is principle precursor** through Krebs (tricarboxylic acid) cycle in which **α-ketoglutarate** is synthesized

(2)   First step in GABA shunt is synthesis of **glutamate** by **transamination of α-ketoglutarate** by enzyme **GABA-α-oxoglutarate transaminase (GABA-T)**; this enzymatic reaction is **coupled to availability of GABA to serve as amine donor**

(3)   **GABA is synthesized from glutamate** by enzyme **glutamic acid decarboxylase (GAD)** which is found only in central nervous system tissue

(4)   Since GABA serves as amine donor for enzyme GABA-T, GABA is only catabolized if its precursor is formed; this provides for conservation of GABA levels

(5)   Deamination of GABA by enzyme GABA-T results in succinic semialdehyde which is then oxidized to succinic acid by enzyme succinic semialdehyde dehydrogenase; succinic acid reenters Krebs cycle

b.   Major pathways

    (1)   Distributed widely in central nervous system in **inhibitory interneurons**

    (2)   **Purkinje cells** — terminals of cerebellar Purkinje cell axons release GABA

    (3)   **Striatonigral GABAergic system** — striatal neurons that have axon projections to midbrain substantia nigra (which contains highest known concentration of GABA)

c.   Catabolism — after release into synaptic cleft, reuptake pumps on presynaptic membrane and on astrocyte membrane (high-affinity uptake sites) transport GABA intracellularly

    (1)   In axon terminals, GABA is available for reutilization

    (2)   Since astrocytes lack glutamic acid decarboxylase, astrocyte GABA is metabolized by GABA-T to succinic semialdehyde and cannot be resynthesized

d.   **GABAergic receptors**

    (1)   **$GABA_A$** — part of **GABA/benzodiazepine receptor/chloride ion channel complex**

        (a)   Stimulation of GABA receptor site produces **increase in mean channel open time** with consequent increased influx of chloride and **hyperpolarization** of postsynaptic membrane

        (b)   Response of receptor to GABA is modified by stimulation of other receptors attached to this complex

            i)   **Benzodiazepine receptor** — stimulation facilitates channel opening in response to stimulation of GABA receptor; drugs interacting at this receptor have tranquilizing and anticonvulsant properties

            ii)   **Barbiturate receptor** — stimulation prolongs channel opening in response to stimulation of GABA receptor; drugs interacting at this receptor have sedating and anticonvulsant properties

    iii)   **Picrotoxin receptor** — stimulation reduces channel opening in response to stimulation of GABA receptor; drugs interacting at this receptor produce generalized convulsions (epileptic seizures)

    (c)   GABA$_B$ — located on presynaptic membrane (autoreceptor); action coupled to G-proteins with inhibition of adenylyl cyclase and consequent inhibition of calcium ion channels (reducing influx of calcium ions and, thereby, inhibiting release of neurotransmitter)

  e.   Pharmacology

    (1)   GABA agonist — muscimol (isolated from hallucinogenic mushroom *Amanita muscaria*) specifically stimulates GABA$_A$ receptors

    (2)   GABA antagonist — **bicuculline** specifically inhibits GABA$_A$ receptor, markedly reducing channel open time; results in clinical convulsions (epileptic seizures)

6. **Glycine**

  a.   **Inhibitory neurotransmitter** synthesized from serine by enzyme serine hydroxymethyltransferase

  b.   Major pathways — inhibitory neurotransmitter of **spinal interneurons** synapsing on anterior horn cells

  c.   Catabolism — after release into synaptic cleft, reuptake pumps on presynaptic membrane (high-affinity uptake sites) transport glycine intracellularly into presynaptic axon terminal

Glycine.

d. **Glycine receptors**

    (1)   Linked to **chloride ion channel** with properties similar to $GABA_A$ receptor

    (2)   Receptor activation increases channel open time for chloride ion passage (thereby, producing membrane hyperpolarization)

    (3)   Receptor can be activated not only by glycine, but also by ß-alanine, taurine, L-alanine, and L-serine; receptor is not activated by GABA

    (4)   **Strychnine** — selective inhibitor of glycine receptor, markedly reducing open time of chloride channel; results in diffuse vigorous muscle spasms

7. **Glutamate** (glutamic acid) and **aspartate** (aspartic acid)

Glutamic acid.

    a.   **Excitatory amino acid neurotransmitters** synthesized in central nervous system primarily from glucose as precursor

    b.   Major pathways — distributed widely in central nervous system; neurotransmitter of cerebellar granule cells

    c.   Catabolism — after release into synaptic cleft, reuptake pumps on presynaptic membrane (high-affinity uptake sites) transport glutamate intracellularly

    d.   **Glutamatergic receptors**

        (1)   **NMDA (*N*-methyl-D-aspartate) receptor**

            (a)   Present throughout cerebral hemisphere (cerebral cortex and basal ganglia), but found in highest concentration in hippocampus

            (b)   Unique receptor complex in that it requires simultaneous binding of two different agonists for activation: glutamate (or aspartate) and glycine

(c) **Ligand-gated ion channel** that when opened allows influx of calcium ions and sodium ions with resultant membrane depolarization; activation of receptors increases channel open time allowing ion movement

(d) Magnesium ion (Mg^{2+}) — blocks open ion channel preventing influx of calcium or sodium ions

(e) Zinc ion (Zn^{2+}) — blocks channel opening

(2) **AMPA** (α-amino-3-hydroxy-5-methyl-4-isoxazolepropionic acid) receptor (**"quisqualate receptor"**)

(a) Distributed widely throughout central nervous system

(b) Mediates fast excitatory synaptic transmission

(c) Stimulated by glutamate; unresponsive to aspartate

(d) **Ligand-gated ion channel** that when opened allows influx of calcium ions and sodium ions with resultant membrane depolarization; activation of receptors increases channel open time allowing ion movement

(3) **Kainate (KA) receptor** — ligand-gated ion channel that when opened allows influx of calcium ions and sodium ions with resultant membrane depolarization; activation of receptors increases channel open time allowing ion movement

(4) Metabotropic receptors — action coupled to G-proteins

(a) L-2-amino-4-phosphonopropionic acid (L-AP4) receptor — found on retinal bipolar neurons; activation of receptor results in membrane hyperpolarization

(b) Aminocyclopentyl dicarboxylic acid (ACPD) receptor — activation of receptor results in activation of phospholipase C and phosphoinositide hydrolysis with subsequent release of calcium ions from intracellular stores

e. **Excitotoxicity** — excessive stimulation of glutamatergic receptors allows **excessive influx of calcium ions** which disrupt cellular metabolism and lead to cell death

  (1) **Ischemia** — following anoxic insult, released glutamate activates receptors leading to calcium ion influx and death of more cells than would have been injured by anoxia alone; experimental administration of NMDA receptor blocker (antagonist) even several hours after anoxic event results in significant protection from death of neurons in hippocampus and striatum

  (2) **Neurolathyrism** — progressive neurodegenerative disease of East Africa associated with dietary consumption of pea *Lathyrus sativus* which contains toxic glutamatergic agonist

8. **Opioids**

a. **Endogenous neurotransmitter peptides** (neuroactive peptides) that **mimic effects of exogenously administered plant alkaloids opium** and morphine

  (1) **Proopiomelanocortin** — precursor protein that can be cleaved into α-endorphin, ß-endorphin, γ-endorphin

  (2) **Proenkephalin** — precursor protein that can be cleaved into Met-enkephalin, Leu-enkephalin

  (3) **Prodynorphin** (proenkephalin B) — precursor protein that can be cleaved into α-neoendorphin, ß-neoendorphin, dynorphin A, dynorphin B

b. Major pathways — distributed widely in central nervous system in many pathways

c. **Catabolism** — after release into synaptic cleft, **peptidases** such as neutral metalloendopeptidase (enkephalinase) or animopeptidase-N cleave neuroactive peptides into inactive fragments

d. Opioid receptors

    (1)   Mu (μ) receptor

        (a)   Agonists — morphine and opiate alkaloids, ß-endorphin, enkephalins

        (b)   Action coupled to G-proteins with inhibition of adenylyl cyclase activity and opening of potassium ion channels

    (2)   Delta (δ) receptor

        (a)   Agonists — enkephalins

        (b)   Action coupled to G-proteins with inhibition of adenylyl cyclase activity and opening of potassium ion channels

    (3)   Kappa (κ) receptor

        (a)   Agonists — dynorphins

        (b)   Action coupled to G-proteins with inhibition of adenylyl cyclase activity and closing of calcium ion channels

e.   Pharmacology

    (1)   Stimulation of opioid receptors results in reduced pain perception (analgesia), ventilatory depression, hypothermia, and altered cardiovascular function

    (2)   Receptor antagonists naloxone or naltrexone reverse effects of exogenous opium or opium analogues

9.   Neuroactive peptides

a.   Peptides colocalize with other (classic) neurotransmitters in axon terminals and when released during synaptic transmission act to modify synaptic transmission by classic neurotransmitter

b.   Representative peptides — **somatostatin**, cholecystokinin (CCK), vasoactive intestinal peptide (VIP), **substance P**, neurotensin

V. **Neuroglia** include macroglia (astrocytes, oligodendrocytes, ependymal cells) and microglia

    A. **Astrocyte** — branched stellate cell with oval nucleus (containing no visible nucleolus), little perinuclear cytoplasm, and numerous cell processes

        1. **Fibrous astrocytes**

            a. Cells with numerous thin, spindly processes; many processes terminate as bulbous expansions (foot plates) which contact walls of blood vessels (**perivascular glial limiting membrane**) or pial surface (**glia limitans**)

            b. Cytoplasm contains dense sheaves of intermediate filaments composed of **glial fibrillary acidic protein (GAAP)**

        2. **Protoplasmic astrocytes** — astrocytes in gray matter closely associated with neurons and serving as metabolic intermediary for neurons

        3. Metabolically active cell

            a. Highly **permeable to potassium ions** (K^+); **buffers extracellular concentration of potassium ions** by taking up potassium ions during intense neuronal activity

            b. Regulates concentrations of free neurotransmitters by taking up, storing, and metabolizing neurotransmitters

            c. Stores and transfers metabolites from capillaries to neurons

    B. **Oligodendrocyte** (oligodendroglia)

        1. Cell with small, darkly-staining, round nucleus, minimal perinuclear cytoplasm, few processes, and no cytoplasmic filaments

        2. **Myelin sheath**

            a. Myelin sheath surrounding myelinated axons formed from oligodendrocyte membrane (in central nervous system) or from Schwann cell membrane (in peripheral nervous system); one oligodendrocyte makes many internodal segments of myelin around many different axons, while one Schwann cell only makes one internodal myelin segment around one axon

b. Consists of fusion of multiple spiral layers of specialized cell membrane that surrounds axon

    (1) Periaxonal space — gap between innermost spiral of myelin sheath and axonal membrane

    (2) Major dense (period) line — fusion of proteins of inner face (cytoplasmic surface) of cell membrane (after exclusion of all cytoplasm) produces dense line

    (3) Interperiod line — closely apposed (but not fused) proteins of outer face of cell membrane; pathologic process affecting myelin often cause separation along interperiod line

    (4) **Node of Ranvier**

        (a) Adjacent segments of myelin along single axon are separated by node of Ranvier; area of axon at node is not covered by myelin

        (b) Lateral loops — myelin sheath adjacent to node of Ranvier (paranodal region) has retained cytoplasm (opening of major dense line)

        (c) Transverse bands — membrane complexes between lateral loops and axonal membrane (axolemma) that seal myelin sheath to axon and demarcate region of axonal excitability

        (d) **Region of axonal excitability** — concentration of voltage-gated sodium channels (necessary for action potential generation) present only in nodal region of axonal membrane

    (5) Schmidt-Lanterman clefts — areas of retained cytoplasm (membrane not compacted to form major dense line) in middle of internodal myelin sheath

    (6) Mesaxon — cytoplasmic tongue connecting myelin sheath to perinuclear oligodendrocyte or Schwann cell cytoplasm; inner mesaxon is adjacent to axon and outer mesaxon is adjacent to outermost myelin layer

        c.   Chemical composition

           (1)   Composed of approximately 75% lipid and 25% protein

           (2)   Major lipids of myelin — **cerebroside**, cholesterol, glycerophospholipids, lecithin, sphingomyelin

           (3)   Protein components (different forms in central and peripheral nervous systems)

                  (a)   **Central nervous system myelin proteins — proteolipid protein (PLP; Folch-Lees protein), myelin basic protein (MBP), myelin-associated glycoprotein (MAG)**

                  (b)   Peripheral nervous system myelin proteins — P_0 **protein, myelin basic protein (P_1), P_2 protein, myelin-associated glycoprotein (MAG)**

C. **Ependymal cell**

    1.   Ciliated cells that form **cuboidal epithelial lining of ventricular cavities**

    2.   Cytoplasmic processes extend deep into parenchyma and contain intermediate filaments (glial fibrillary acidic protein)

    3.   **Choroid plexus epithelium** — modified ependymal cells that cover choroid plexus and produce cerebrospinal fluid

VI. **Microglial** cell — elongated cell with cylindrical nucleus; resting macrophage of central nervous system

VII. **Endothelial cells**

    A. Cells which line blood vessels and separate neural parenchymal from blood compartment

    B. Blood-brain barrier — tight junctions between adjacent endothelial cells exclude most molecules from entry into central nervous system

    C. Contain specialized transport systems which facilitate passage of necessary molecules into central nervous system

VIII. **Pia-arachnoid**

    A.   Delicate membrane surrounding brain and spinal cord

    B.   **Pia** — single layer of cells that is tightly adherent to glial processes (glia limitans) and follows contour of brain and spinal cord

    C.   **Arachnoid**

        1.   Layer of cells separated from pia by space (subarachnoid space) that contains cerebrospinal fluid

        2.   Strands of arachnoidal cells (arachnoid trabeculae) pass between pia and arachnoid and invest structures (such as blood vessels) within subarachnoid space

        3.   Follows general contours of brain and spinal cord, bridging sulci and other areas resulting in focal enlargements of subarachnoid space (subarachnoid cisterns)

    D.   **Virchow-Robin space** — space between arachnoid sheath covering blood vessels that dive into parenchyma and pial barrier of brain parenchyma; continuous with subarachnoid space and can become distended with material (such as purulent exudate, toxins, malignant cells) that fills subarachnoid space

SUGGESTED ADDITIONAL READING

Burgoyne RD: *The Neuronal Cytoskeleton*, New York, Wiley-Liss, 1991

Cooper JR, Bloom FE, Roth RH: *The Biochemical Basis of Neuropharmacology*, 6th ed. New York, Oxford University Press, 1991.

Nolte J: *The Human Brain*, 3rd ed. St. Louis, Mosby, 1993.

Siegel GJ, Agranoff BW, Albers RW, Molinoff PB (eds): *Basic Neurochemistry*, 5th ed. New York, Raven Press, 1994.

CHAPTER 2: PRINCIPLES OF CEREBRAL LOCALIZATION

I. Each nervous system area has specific functions and circumscribed (focal) nervous system lesions produce characteristic clinical signs and symptoms; neurologic and neurophysiologic studies can identify these signs and symptoms and suggest appropriate neuroanatomic localization for lesions

   A. **Hemisyndromes** — symptoms confined to only **one side of body** are produced by disorders of cerebral hemispheric motor or sensory pathways

   B. **Behavioral syndromes** (disorders involving "**higher cortical functions**") — aphasias, apraxias, agnosias, and similar behavioral syndromes (sometimes with other motor or sensory signs) result from lesions of circumscribed cerebral cortical areas

   C. **Frontal lobe syndromes** — localized contralateral paralysis, focal motor epileptic seizures, reappearance of primitive reflexes (such as rooting reflex which is normal in infants, but disappears with maturation), involuntary resistance to passive limb movement, perseveration (abnormal persistence or repetition), apathy, motor aphasia (lesion localized to Broca's area), inappropriate or uninhibited social behavior

   D. **Parietal lobes syndromes** — contralateral sensory hemisyndromes, agnosia (inattention to opposite visual field or side of body), abnormal tactile discrimination, apraxia (difficulty performing purposeful or complex movement without evidence of paralysis or sensory loss)

   E. **Temporal lobe syndromes** — contralateral homonymous visual field defects (particularly upper quadrantanopia due to disruption of visual radiations in Meyer's loop), receptive language disorders (such as Wernicke's aphasia), memory disturbances, epileptic seizures (complex partial seizure disorder)

F. **Occipital lobe syndromes** — contralateral visual field defects, alexia (inability to read), color recognition disturbance

G. **Basal ganglia syndromes (extrapyramidal disorders)**

    1.   Loss of neurons in substantia nigra or globus pallidus — hypokinesia (reduced movement), lack of associated or autonomic movements

    2.   Loss of neurons in striatum — hyperkinesia (increased involuntary movements), choreoathetosis

H. **Thalamic syndromes** — impairment of contralateral sensation (delayed sensation, unpleasant or painful sensations), abnormal contralateral posturing (particularly of hands)

I. **Hypothalamic syndromes** — impaired regulation of temperature, salt and water metabolism, feeding, and sleep-wake cycle, altered sexual functioning, endocrine disturbances

J. **Brain stem syndromes** — "crossed deficits" ("nuclear" motor or sensory deficits on ipsilateral side of head or face and "central" deficits on contralateral side of body), impaired ocular motility, nystagmus, vertigo, altered consciousness

K. **Cerebellar syndromes** — ataxia or dysmetria, intention tremor, hypotonia, nystagmus

L. **Spinal cord syndromes**

    1.   Transection — spastic paralysis of extremities below level of lesion, loss of sensation below level of lesion (sensory level), paralysis of rectal sphincter function, abnormal bladder function, disturbed sexual function (in males)

    2.   Brown-Séquard syndrome (spinal cord hemisyndrome) — **ipsilateral spastic weakness, ipsilateral loss of proprioception** (vibration and position sense), and **contralateral loss of pain and temperature** sensation below level of lesion

    3.   Centromedullary infarction (**anterior spinal artery syndrome**) — **paralysis** and **dissociated sensory loss (loss of pain and temperature** sense, but **preservation of proprioception**) below level of lesion

36

4. Anterior horn destruction — flaccid paralysis with muscle fasciculations, fibrillation, and atrophy

M. Peripheral nervous system syndromes — dermatomal sensory loss and flaccid paralysis

II. **Visual disturbances** — characteristic visual field losses indicate site of lesion along pathway from optic nerve to occipital cortex

III. Vertigo, dizziness, unsteadiness

A. **Vertigo** (sensation of spinning) indicates dysfunction of peripheral vestibular apparatus, vestibulocochlear nerve (cranial nerve VIII), or brain stem vestibular nuclei

B. **Sensory ataxia** — dizziness or unsteadiness secondary to impaired sensory input (as in peripheral neuropathies) is worsened by darkness, eye closure, or rapid head movements

C. **Cerebellar ataxia** — unsteadiness associated with hypotonia, intention tremor, and truncal instability

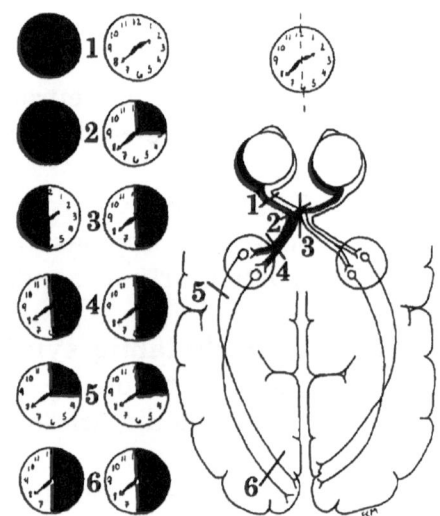

Visual field defects associated with lesions in various parts of visual system.

IV. **Unilateral facial paralysis**

A. **Peripheral facial paralysis**

1. Lesion of facial nerve (cranial nerve VII) or facial nucleus

2. All ipsilateral facial muscles (including forehead and periorbital muscles) equally involved, incomplete eye closure (sclera exposed) with visible reflex upward movement of eyes during attempted eye closure (Bell's phenomenon), disturbed taste on anterior two-thirds of tongue, impaired tearing

B. **Central facial paralysis**

1. Lesion damaging corticobulbar fibers terminating in facial nucleus; often accompanied by other hemisyndromes

2.   Facial nerve branches to frontalis and periorbital muscles less affected than rest of facial nerve, resulting in less paralysis of these muscles; eye closure covers sclera

V.   Gait disturbance

A.   **Steppage gait — foot extensor weakness** necessitates excessive elevation of legs to clear toes above ground in swinging leg forward and then toes strike (often slapping) ground first

B.   **Spastic gait** — with corticospinal tract damage, feet drag with little knee movement, leg is often internally rotated with foot turned inward (pigeon toes), and slightly bent knee crosses midline with each step forward (scissoring)

C.   **Parkinsonian gait** — stiff, slow movements, posture slightly flexed forward with knees and elbows flexed, small steps

D.   **Ataxic gait — wide-based, unsteady movement** with abrupt irregular placement of feet (stamping of feet) and swaying of trunk

E.   **Paretic gait — quadriceps paralysis** necessitates locking of knee (knee hyperextension) on supporting leg to prevent collapse

VI.   Epilepsy and seizure disorders

A.   **Seizures**

1.   Abnormal behavior resulting from aberrant spontaneous neuronal electrical discharge; results from abnormal hypersynchronized activation of large population of cerebral neurons

2.   Can be induced (provoked) in otherwise normal nervous system under certain circumstances such as hypoglycemia, administration of convulsant drugs, or with electrical stimulation

3.   **Partial seizures** — seizures that begin focally, such as focal motor movements (jerking) or alterations in behavior or

Example of recording of two channels of EEG activity.

consciousness; complex partial seizures have complicated (complex) behavior and motor features

4. **Generalized seizures** — bilaterally symmetric seizures that involve whole cerebral cortex nearly simultaneously (no evidence of focal onset)

   a. **Tonic-clonic (grand mal) seizures** — bilateral tonic (stiffening) and clonic (rhythmic jerking) of nearly all muscles, loss of consciousness, tongue biting, and urinary incontinence

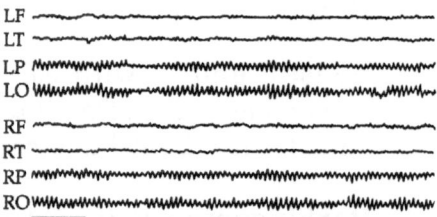

Normal EEG tracing showing alpha rhythm in parietal and occipital leads (marker indicates 1 second).

   b. **Absence (petit mal) seizures** — brief episodes of staring (sometimes with associated eye blinking)

B. **Epilepsy** — tendency to have recurrent unprovoked seizures

C. **Electroencephalography (EEG)** — electrophysiological recording technique used to identify cerebral electrical activity associated with seizures

   1. Recording of surface cerebral cortical electrical activity (potentials) using multiple electrodes placed over scalp (or exposed brain following neurosurgical procedures)

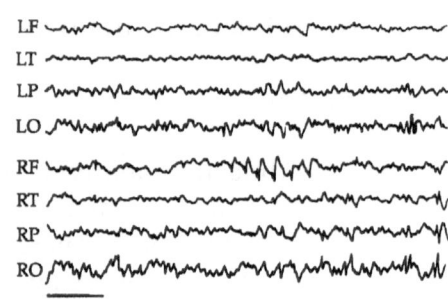

EEG in complex partial seizure disorder showing spikes in multiple leads.

   2. Frequencies of activity recorded vary from 0.5-50 Hz; amplitude of activity ranges from 1-100 μV

   3. Recorded activity usually appears in several frequency ranges:

   a. **Alpha rhythm** — rhythmic 8-13 Hz activity recorded from parieto-occipital region during relaxed wakefulness with eyes closed

b.   Beta activity — low amplitude rhythmic 14-30 Hz ("fast") activity most prominent in frontal region and accentuated by psychoactive medications (particularly sedatives)

c.   Theta activity (4-7 Hz) and delta activity (0.5-4 Hz)

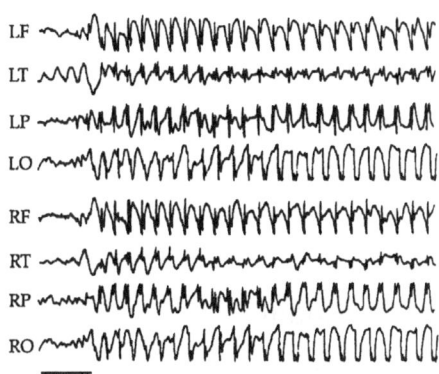

EEG showing 3 Hz spike-and-wave pattern of absence seizures.

    (1)   Bilateral regular rhythmic "slow wave" activity occurs normally during deeper stages of sleep

    (2)   Unilateral slow wave activity occurs over regions of cerebral abnormality (such as stroke or tumor)

    (3)   Bilateral slow wave activity occurs with diffuse cerebral disorders associated with altered level of consciousness such as metabolic derangements, diffuse edema, brain stem lesions

4.   **Spikes** — high amplitude, high frequency potentials that **characterize seizures**; similar potentials that have slightly lower frequency are termed "sharp waves"

SUGGESTED ADDITIONAL READING

Brodal P: *The Central Nervous System: Structure and Function.* New York, Oxford University Press, 1992.

Mumenthaler M: *Neurologic Differential Diagnosis*, second edition. New York, Thieme Medical Publishers, 1992.

Plum F, Posner JB: *The Diagnosis of Stupor and Coma*, edition 3. Philadelphia, F. A. Davis Company, 1980.

CHAPTER 3: BRAIN EDEMA AND MASS EFFECT

I. **Edema**

    A. **Brain edema** – increase in brain volume due to increased tissue water content

    B. **Vasogenic edema** – extravasation of plasma into brain parenchyma following damage to blood-brain barrier

    C. **Cytotoxic edema** – increased intracellular water (primarily in gray matter astrocytes) usually following hypoxic-ischemic injury

    D. **Interstitial edema** – increased water content in periventricular white matter associated with hydrocephalus (hydrostatic edema)

II. Intracranial fluid dynamics

    A. Intracranial volumes

        1. Total intracranial capacity — 1600 cc

        2. Brain volume — 1200-1500 cc (based on normal weight range of 1200-1500 gm and approximate conversion of 1 gm/cc)

        3. Intravascular (blood) volume — 75 cc

        4. Cerebrospinal fluid (ventricular and subarachnoid) volume — >75 cc

    B. **Cerebrospinal fluid**

        1. Formed intraventricularly at rate of 20 mL/hour (480 mL/day)

2. Normally contains less than 5 white blood cells per microliter, all mononuclear cells (no neutrophils)

3. Pressure

    a. Normal pressure ranges from 50-200 mm of water (values below 180 mm are clearly normal; values above 200 mm are clearly elevated)

    b. Fluctuations, or pressure waves, secondary to pulse and respirations are normally present

    c. Type A or plateau (Lundberg) waves — prolonged and marked elevations of pressure occurring in pathologic states

C. Intracranial cavity has limited space for tissue expansion — any change in one intracranial component requires compensatory changes in other components and alters intracranial pressure

1. Increase in cerebrospinal fluid volume (as in hydrocephalus) necessitates compensatory reduction in brain parenchymal volume

2. Increase in brain parenchymal volume (as in cerebral edema) necessitates compensatory reduction in cerebrospinal fluid or vascular (blood) volume

    a. Sufficient increase in brain volume can compress vessels enough to eliminate all blood flow (brain death)

    b. Herniation — once intracranial capacity is exceeded, brain parenchyma will be extruded through cranial openings such as foramen magnum

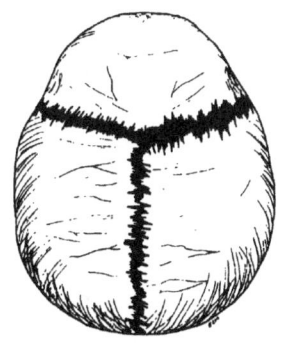

Splitting of cranial sutures with increased intracranial pressure in children

3. Open bony sutures in infants and children and cerebral atrophy in elderly provide extra space for tissue expansion

III. **Brain herniation** — displacement of brain tissue from one intracranial compartment to another or through skull foramina

A. **Intracranial compartments**

1. **Infratentorial compartment** (posterior fossa) lies beneath tentorium cerebelli

2. **Supratentorial compartments**

   a. Left and right are separated by falx cerebri

   b. Anterior and middle cranial fossae on each side

B. Types of herniation

1. **Subfalcine herniation**

   a. Unilateral cerebral hemisphere mass or edema (usually frontal or frontoparietal) produces displacement of medial cerebral hemisphere (usually cingulate gyrus) under falx cerebri

   b. Can produce occlusion of anterior cerebral artery branches, with resultant infarction of supplied medial cerebral hemisphere tissue

2. **Transtentorial herniation** (herniation through tentorial incisura)

   a. **Most common form** of cerebral herniation

   b. Unilateral cerebral mass lesion or edema causes **herniation of ipsilateral medial temporal lobe** (particularly uncus) into infratentorial compartment

      (1) Compression of **ipsilateral oculomotor nerve (cranial nerve III)** — results in pupillary dilation

      (2) **Kernohan's notch**

         (a) **Incision of opposite cerebral peduncle by sharp edge of tentorium**

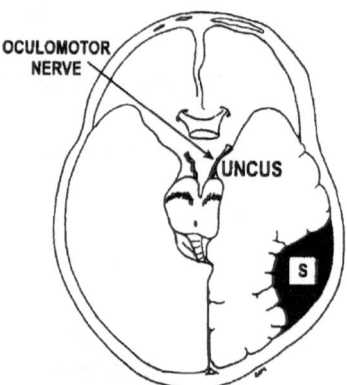

Mass effect of temporal subdural hematoma causes trantentorial uncal herniation with compression of midbrain and oculomotor nerve.

    (b)   Clinical symptoms — **hemiplegia contralateral to incised cerebral peduncle, but ipsilateral to herniating cerebral hemisphere**

(3)   Distortion and elongation of midbrain and of cerebral aqueduct of Sylvius ("slit-like" aqueduct)

(4)   **Duret hemorrhage** — rupture of midline penetrating brain stem blood vessels resulting in midbrain hemorrhage and death

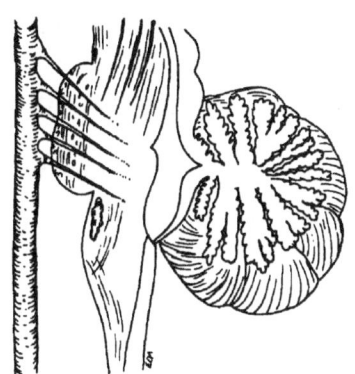

(5)   Compression of posterior cerebral artery and unilateral infarct of visual cortex

Stretching and tearing of penetrating vessels in pons during herniation results in Duret hemorrhages.

c.   Diffuse bilateral cerebral mass lesion or edema causes herniation of bilateral medial temporal lobes into infratentorial compartment

(1)   Diencephalic (thalamic) dysfunction — disruption of reticular activating system results in alteration of consciousness

(2)   Compression of blood vessels of circle of Willis

    (a)   Results in cerebral ischemia

    (b)   Bilateral occlusion of posterior cerebral arteries produces bilateral infarcts of occipital cortex (visual cortex)

(3)   Bilateral compression of midbrain (particularly cerebral peduncles) — results in bilateral hemiplegia (decerebrate posturing)

(4)   Duret hemorrhage

3.   **Upward cerebellar herniation** — diffuse cerebellar mass lesion or edema results in upward herniation of midline vermis through tentorial incisura

   a.   Compression of midbrain aqueduct produces obstructive hydrocephalus

   b.   Compression of superior cerebellar arteries produces infarcts of superior surface of cerebellum

4.   **Cerebellar tonsillar (downward) herniation**

   a.   Occurs with either supratentorial or subtentorial mass lesions or edema

   b.   **Tight cone of cerebellar tissue around medulla** presses down into (herniates through) foramen magnum compressing entire brain stem and obliterating fourth ventricle

      (1)   **Cushing reflex — medullary compression** causes **irregular shallow breathing** (ataxic breathing), elevation of systolic blood pressure with **widened pulse pressure** (marked elevation of systolic pressure with minimal elevation or reduction in diastolic pressure), and **bradycardia (slowed heart rate)**

      (2)   Infarction of compressed cerebellar tonsils results in cerebellar tissue being sloughed into spinal subarachnoid space

SUGGESTED ADDITIONAL READING

Plum F, Posner JB: *The Diagnosis of Stupor and Coma*, edition 3. Philadelphia, F. A. Davis Company, 1980.

CHAPTER 4: PATHOLOGY OF CEREBROVASCULAR DISEASE

I.   **Stroke**

    A.   **Sudden catastrophic neurologic event**

    B.   Includes cerebral **arterial thrombosis** or **embolus, intracerebral hemorrhage, primary subarachnoid hemorrhage,** or **transient ischemic attack (TIA)**

    C.   Third leading cause of death in United States for individuals over age 65 years; approximately 500,000 strokes per year

II. **Hypoxia, ischemia, infarction**

    A.   **Hypoxia** (or anoxia)

        1.   **Insufficient oxygen to meet tissue requirements**

            a.   Brain energy supply derived almost entirely from oxidative metabolism of glucose

            b.   Brain makes up only 2% of body weight, but brain metabolism normally accounts for 20% of total body oxygen consumption

        2.   **Anoxic hypoxia** — reduction of oxygen delivery to tissues due to **decreased oxygen saturation of circulating blood** (example: asphyxia)

        3.   **Anemic hypoxia** — reduction of oxygen delivery to tissues due to **decrease in available hemoglobin to bind oxygen** (example: following massive hemorrhage)

B. **Ischemia**

    1.   **Insufficient blood supply to meet tissue requirements** — brain normally requires 20% of cardiac output

    2.   Usually results from vascular obstruction or impairment of cardiac output (such as in cardiac arrest)

C. **Infarction**

    1.   **Necrosis resulting from sudden reduction in oxygen delivery, arterial blood supply, or venous drainage of brain**

    2.   Pathologic features

        a.   **24 hours** — circumscribed area of tissue swelling, softening, and discoloration; blurring of structural details (such as gray matter-white matter junction); **cytoplasmic eosinophilia** and **nuclear pyknosis** of neurons and other cellular elements (coagulative necrosis)

        b.   **48 hours** — necrotic area becomes mushy and friable and clearly demarcated from viable tissue; neutrophilic infiltration beginning at edge of necrotic area

        c.   **96 hours** — **maximal swelling (edema)** from breakdown of blood-brain barrier in both necrotic area and surrounding tissue; axonal retraction balls (axonal swellings) at edge of necrotic area from disruption of axons that would normally pass through necrotic area; blood monocytes infiltrating tissue at margin of necrotic area are transformed into lipid-laden macrophages ("gitter cells")

        d.   **7 days** — **capillary hyperplasia** and **astrocytic proliferation** and hypertrophy ("gemistocytic astrocytes") at margin of necrotic area

        e.   **14 days** — **liquifaction** of necrotic tissues; sheets of lipid-laden and hemosiderin-laden macrophages throughout necrotic area

        f.   **21 days** — cystic space (**cavitation**) containing numerous macrophages and surrounded by rim of hypertrophic astrocytes

g. **3 months — fluid filled cystic space** which is traversed by scattered vessels, glial processes, and collagen fibrils; occasional macrophages persist

h. Note:

   (1) Molecular layer survival — in cerebral cortical infarcts, molecular layer (layer I) usually survives despite otherwise complete tissue necrosis

   (2) Neonatal infarction — tissue dissolution is more rapid, presumably due to relative lack of myelin lipids

   (3) **Incomplete infarction** — only neurons and oligodendrocytes die while astrocytes and capillaries survive resulting in densely gliotic, slightly rarefied tissue, but no cavitation

   (4) **Hemorrhagic infarction — leakage of blood into infarcted tissue**, usually resulting from delayed restoration of circulation after tissue necrosis has occurred (as with vascular obstruction by embolus which then breaks apart allowing blood flow to resume); large amounts of hemosiderin (brown) and hematoidin (yellow) from blood breakdown impart brownish-yellow color to cavity

D. **Cerebral autoregulation**

1. **Blood flow to brain remains constant despite variation in perfusion pressure** (blood pressure) due to intrinsic tendency for arteriolar dilation or constriction

2. Acute brain damage can disrupt autoregulatory mechanism and thus ability to maintain cerebral blood flow under conditions of fluctuating blood pressure

E. **Selective vulnerability** — greater vulnerability of certain brain cells and areas to hypoxia or ischemia; most vulnerable to destruction include:

1. **Neurons > oligodendrocytes > astrocytes > endothelial cells**

2. **Cerebral neocortical layers III, V, VI**

3. **Cerebellar Purkinje cells**

4.   **Cerebral cortex at depths of gyri**

5.   **Sommer's sector of hippocampus**

6.   Parietal and occipital lobes > frontal and temporal lobes

7.   Outer (lateral) half of caudate nucleus and putamen

8.   Anterior and dorsomedial thalamic nuclei

9.   Substantia nigra and inferior colliculi

F.   **Risk factors for stroke — hypertension, diabetes mellitus, cardiac disease** (myocardial infarction, rheumatic heart disease, **atrial fibrillation**, congenital heart disease, subacute bacterial endocarditis), **systemic hypotension**, hematologic disorders (polycythemia, **sickle cell disease**), autoimmune disorders (systemic lupus erythematosus, antiphospholipid antibody syndrome), cerebrovascular structural anomalies (**fibromuscular dysplasia, arteriovenous malformation**), trauma

III. **Hypoxic/ischemic encephalopathy**

A.   **Brain damage resulting from reduction in brain perfusion or oxygen content of blood flowing to brain**

B.   **Ischemic cell change – neuronal eosinophilia (eosinophilic cytoplasm) and nuclear pyknosis** developing within 24 hours of hypoxia/ischemia

C.   **Morphologic patterns** — variability based upon severity of hypoxia, rapidity of onset, condition of collateral vascular supply, relative susceptibility of cellular elements, and associated metabolic factors (such as acidosis, hypoglycemia)

   1.   Patterns occurring at any age:

      a.   **Watershed (or border zone) infarction**

         (1)   Necrosis of brain in **junctional areas between vascular supply** (territories) of major arteries

         (2)   Usually associated with **hypotension** and consequent reduced perfusion at most distant extreme of vascular territory

b.   **Laminar cortical necrosis**

    (1)   Extensive loss of neurons in cerebral cortical layers III, IV, V; results in **slit cavity paralleling gray matter-white matter junction**

    (2)   Usually associated with total (global) ischemia (as occurs with **cardiac arrest**)

c.   **Sulcal infarcts**

    (1)   **Necrosis of cerebral cortex in depths of sulci**

    (2)   Usually associated with hypotension complicated by cerebral edema (swelling) in which swollen gyri compress arteries nourishing cortical tissue at depths of sulci

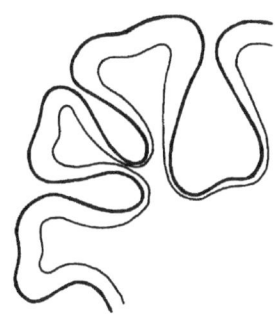

    (3)   **Ulegyria** — gyri have mushroom shape (cerebral cortex at crests of gyri is thicker than at depths of sulci) following sulcal infarction (particularly in children and most often involving frontal or occipital lobes)

Ulegyria: mushroom-shaped gyri due to greater destruction at depth of sulcus.

d.   **Granular atrophy**

    (1)   **Pitted appearance of cerebral cortical surface** resulting from columns of necrotic cortex which appear sunken next to normal cortex; results from **hyaline arteriolosclerosis** and occlusion of penetrating arterioles

    (2)   Often associated with **labile hypertension in elderly individuals**

e.   **Necrosis of Sommer's sector of hippocampus**

    (1)   **Necrosis of pyramidal neurons in area of hippocampus (Sommer's sector)** supplied by endarteries

(2)   Commonly associated with all types of hypoxic/ischemic episodes

2.   Patterns occurring in infants

    a.   **Status marmoratus**

        (1)   **Loss of neurons** in basal ganglia (caudate and putamen) and thalamus with subsequent **abnormal myelination** of astrocytic processes by surviving oligodendroglia cells

        (2)   Associated with perinatal hypoxic/ischemic damage

    b.   **Kernicterus**

        (1)   Immature neonatal blood-brain barrier damaged by perinatal hypoxia/ischemia permits **unconjugated bilirubin** to enter brain resulting in varying degrees of **bilirubin deposition** (yellow discoloration) of basal ganglia, thalamus, hippocampus, inferior olivary nuclei, subthalamic nuclei, cerebellar dentate nuclei, and most cranial nerve nuclei

        (2)   In full term neonates, usually associated with **markedly elevated serum bilirubin levels** of **erythroblastosis fetalis** (hemolysis due to maternal-fetal blood group or Rh incompatibility)

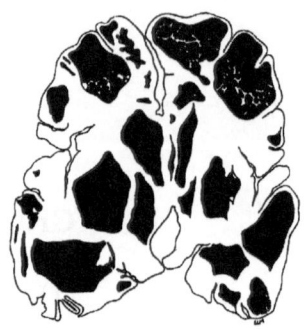

Multicystic encephalopathy.

        (3)   In premature neonates, slight elevation of serum bilirubin levels may result in brain bilirubin deposition

    c.   **Periventricular leukomalacia**

        (1)   Foci of **necrosis in deep periventricular white matter** particularly near foramen of Monro and occipital lobe

        (2)   Associated with hypoxia/ischemia in premature neonates

    d.   **Multicystic encephalopathy**

        (1)   **Widespread cystic necrosis** of cerebral cortex, white matter, basal ganglia, and thalamus

        (2)   Associated with severe in utero (or rarely perinatal) hypoxia/ischemia

D.   **Brain infarcts**

    1.   Causes include

        a.   **Atherosclerosis** — disease of **large arteries** (particularly carotid and basilar arteries) characterized by **atheromatous plaque** formation (focal thickening of intima due to lipid deposition and fibrosis), calcification, ulceration, thrombus formation, hemorrhage, and aneurysmal dilation

        b.   **Arteriolosclerosis** — disease of **arterioles** characterized by vessel wall fibrosis, **hyalinization** (lipohyalinosis), thickening (with occlusion of lumen), and destruction (resulting in hemorrhage); usually associated with **hypertension**

        c.   **Emboli**

            (1)   Intravascular solid, liquid, or gaseous mass carried by blood to site distant from its site of origin

            (2)   Common sources

                (a)   Thrombus in atrial appendage associated with atrial fibrillation

                (b)   Aortic or mitral valvulitis — bacterial endocarditis or nonbacterial thrombotic (marantic) endocarditis

                (c)   Cardiac (atrial) myxoma

                (d)   Atheromatous material from ulcerated plaques in ascending aorta or carotid artery (particularly at carotid bifurcation)

                (e)   Fat — particularly following comminuted long bone fractures

(f)   Carcinoma — particularly lung carcinoma

d.   **Arteritis** — inflammation of walls of arteries or arterioles resulting in blockage of vessel lumen

2.   **Internal carotid artery occlusion**

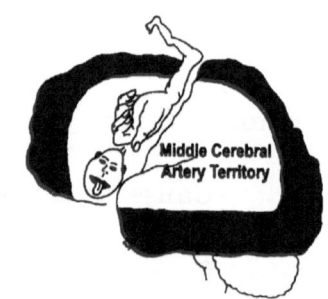

Territory of middle cerebral artery.

a.   Common result of atherosclerosis with **thrombosis at origin of internal carotid artery** (carotid bifurcation)

b.   Ischemia distal to stenosis does not occur until arterial lumen is reduced by greater than 80%

c.   Infarction most commonly involves **middle cerebral artery territory** of cerebral hemisphere (entire lateral surface of cerebral hemisphere and striatum), but extent of brain infarction dependent on degree of integrity of vascular anastomoses in circle of Willis

3.   Infarction in vertebral-basilar artery territory

a.   Vertebral artery occlusion is more common than basilar artery occlusion

b.   Vertebral artery occlusion is most commonly caused by atherosclerosis, but occlusion can also follow vertebral artery compression by cervical vertebral osteoarthritis, cervical manipulation, or vertebral destruction by rheumatoid arthritis

c.   **Wallenberg syndrome** (lateral medullary syndrome) — unilateral occlusion of vertebral artery or posterior inferior cerebellar artery (branch of vertebral artery) with consequent infarction of dorsolateral medulla

4.   **Middle cerebral artery** occlusion

a.   Commonly results from **embolus** lodging at origin of middle cerebral artery

b.   Infarction usually involves **entire lateral surface of cerebral hemisphere and striatum**, but extent of infarction varies depending upon degree of anastomoses with branches of other cerebral arteries

    5.   Posterior cerebral artery occlusion — results from embolus lodging at top of basilar artery or from compression of posterior cerebral artery by transtentorial uncal herniation

IV. Transient ischemic attacks (TIA's)

A.   **Transient neurologic events indicative of temporary brain ischemia,** but fully reversible within minutes (never longer than 24 hours) with no neurologic sequelae

B.   Considered acute neurologic emergency since these episodes herald ischemic events producing permanent damage

V. Fibromuscular dysplasia

A.   Abnormality of **small to medium-sized arteries** (usually **renal** and **craniocervical**) characterized by **areas of vessel wall thickening (constriction) alternating with areas of dilation** resulting in radiographic "string-of-beads" or "tubular stenosis" appearance

B.   Women affected twice as often as men with clinical symptoms of hypertension and ischemia

C.   Pathologic features consist of segmental discontinuities in smooth muscle and internal elastic lamina (leading to vessel dilation) alternating with fibroblastic proliferation (leading to vessel constriction)

VI. Moyamoya disease

A.   **Spontaneous occlusion of branches of vessels of circle of Willis** associated with development of **abnormal vascular network** producing angiographic appearance of "puff of smoke" (Japanese term *moyamoya*)

B.   Pathologic features

    1.   Infolding and duplication of internal elastic lamina with concentric intimal thickening resulting in stenosis or occlusion

    2.   Network of abnormal thin-walled vessels (representing collateral channels) which are prone to rupture with consequent intraparenchymal hemorrhage

VII. **Cerebral amyloid angiopathy (congophilic angiopathy)**

    A.   **Amyloid deposition** in media and adventitia of leptomeningeal and parenchymal arteries and arterioles

    B.   Thickening of vessel walls results in stenosis or occlusion

    C.   Cerebral **lobar hemorrhage** — degeneration of affected vessel walls results in aneurysmal dilation with subsequent rupture and hemorrhage (often multifocal or recurrent)

    D.   Pathologic features

        1.   Infiltration of vessel walls with **amorphous pink material** that displays green dichroism when stained with Congo red stain and viewed with polarized light

        2.   **Frequently accompanies Alzheimer's disease**

        3.   **Unrelated to systemic amyloidosis**; familial forms are due to **genetic defect in amyloid precursor protein** (APP) leading to deposition of βA4-amyloid

VIII. **Arteritis (vasculitis)**

    A.   Noninfectious vasculitis (collagen vascular or connective tissue disease)

        1.   **Polyarteritis nodosa**

            a.   **Segmental necrotizing panarteritis** involving mainly skeletal muscle and peripheral nerve; central nervous system involvement occurs in about 10% of cases

            b.   Pathologic features consist of infiltration of vessel wall by **neutrophils**, eosinophils, and lymphocytes with resultant vessel wall destruction, thrombosis, and fibrosis

            c.   Involvement of renal vessels results in hypertension which can lead to intracranial hemorrhage or hypertensive encephalopathy

        2.   **Systemic lupus erythematosus (SLE)**

            a.   Neurologic involvement due to vascular damage; particularly prominent in those patients with **antiphospholipid antibodies**

b. Pathologic features consist of ischemic lesions in brain and peripheral nerve

c. Lupus cerebritis — multiple embolic cerebral infarcts from emboli originating on heart valves (Libman-Sacks endocarditis)

3. **Temporal (cranial) arteritis (giant cell arteritis)**

a. **Inflammatory disorder** of extracranial arteries in **older individuals** (over age 50 years) presenting with headache, malaise, generalized muscle and joint aches (**polymyalgia rheumatica**), **thickened, tender, non-pulsatile temporal or other scalp arteries**, and **visual disturbances** (blurring or blindness)

b. Characteristically associated with **marked elevation of erythrocyte sedimentation rate** (ESR), often up to 120 mm/hour ("ESR greater than patient's age")

c. Pathologic features consist of **granulomatous (giant-cell) inflammation** of vessel walls with **fragmentation of elastic lamina** (particularly external elastic lamina) with thickening of wall and consequent stenosis or occlusion of vascular lumen

   (1) Both internal and external elastic lamina present in extracranial vessels, but intracranial arteries only have internal elastic lamina

   (2) Sharp demarcation of inflammation at site of vessel entry into cranial cavity has suggested that this is autoimmune disease directed against external elastic lamina

d. **Temporal artery biopsy** is diagnostic

B. Infectious vasculitis

1. **Meningovascular syphilis — arteritis** produced by **neurosyphilis**, usually occurring 5-10 years after initial infection and producing vascular occlusion with ischemic necrosis of brain

2. **Bacterial or mycotic aneurysm**

    a. Infectious arteritis initiated by septic embolus with subsequent direct invasion by fungi or bacteria resulting in necrotizing inflammation and aneurysm formation

    b. Common inciting organisms are **Streptococcus viridans** or **Staphylococcus aureus** (in bacterial endocarditis) or angioinvasive fungi (in immunocompromised individuals)

IX. **Hypertensive cerebrovascular disease**

A. **Hypertensive encephalopathy** — headache, confusion, altered consciousness, and seizures associated with sudden rapid elevation of blood pressure

B. **Hyaline arteriolosclerosis**

    1. Vascular change affecting small arteries and arterioles mainly in brain and kidney

    2. Pathologic changes progress from hypertrophy of vascular smooth muscle, fibrous replacement (hyalinization) of smooth muscle, breakdown of elastic lamina, tortuous elongation of vessels, vessel wall necrosis with thinning, blood leakage, aneurysm formation, and intra-parenchymal hemorrhage

    3. **Lacunae (lacunar infarct)**

        a. Small round to oval cavities (3 to 20 mm in diameter) in **basal ganglia, deep cerebral white matter, internal capsule,** or **basis pons** associated with hypertension and hyaline arteriolosclerosis

        b. État lacunaire (lacunar state) — multiple small lacunar infarcts in basal ganglia and thalamus

        c. État criblé (cribriform state) — hyalinized cerebral white matter arterioles surrounded by dilated perivascular spaces often containing hemosiderin-laden macrophages

    4. **Binswanger's disease (subcortical arteriosclerotic encephalopathy)**

        a. Clinical syndrome of dementia characterized by progressive deterioration of psychomotor functioning

      b.   Diffuse or patchy loss of white matter (leukoaraiosis — radiographic lucency of white matter)

      c.   Accompanied by lacunar infarcts and severe hyaline arteriolosclerosis

5.   **Miliary aneurysms** (microaneurysms; Charcot-Bouchard aneurysms)

      a.   **Outpouchings of walls of small arteries and arterioles** associated with **hyaline arteriolosclerosis**; often surrounded by hemosiderin-laden macrophages and gliotic brain tissue

      b.   Most often found in basal ganglia and thalamus

      c.   Associated with **spontaneous intraparenchymal hemorrhage**

          (1)   Frequency of sites of intraparenchymal hemorrhage: basal ganglia > cerebral white matter > cerebellum (dentate nucleus) > thalamus > brain stem

          (2)   Small hemorrhages are associated with headache, seizures, and focal neurologic deficits; large hemorrhages result in rapidly progressive coma and death secondary to intraventricular rupture or brain stem compression from transtentorial uncal herniation or cerebellar tonsillar herniation

X.   Intracranial vascular aneurysms

    A.   Types of aneurysms — saccular (berry) aneurysm, atherosclerotic fusiform aneurysm, post-traumatic aneurysm, dissecting aneurysm, mycotic aneurysm, miliary (Charcot-Bouchard) aneurysm, vein of Galen aneurysm

    B.   **Saccular aneurysm**

      1.   Outpouchings at major intracranial arterial branch points; present in approximately 1% of population and often multiple

      2.   **Etiology**

          a.   Intracranial arteries lack external elastic lamina; internal elastic

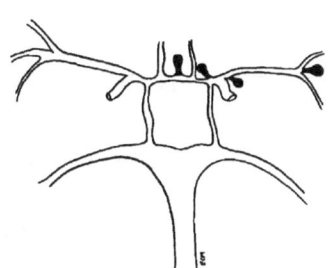

Saccular aneurysms of circle of Willis.

lamina and smooth muscle of media normally provide arterial wall strength

b. Congenital defect of smooth muscle at arterial branch points results in weakened wall with only internal elastic lamina remaining to provide wall strength

c. Hemodynamic stress (increasing with age) damages elastic lamina and results in outpouching that progressively increases in size to form saccular aneurysm

3. **Symptoms and complications**

a. **Mass effect of large aneurysm compresses adjacent structures**

b. Increases in hemodynamic stress (such as increases in blood pressure) on aneurysm wall can result in **rupture** with consequent **subarachnoid or intraparenchymal hemorrhage**

c. **Arterial vasospasm**

(1) **Marked arterial constriction following aneurysm rupture and subarachnoid hemorrhage**, resulting in brain ischemia and **infarction**

(2) Commonly occurs several days (most severe at seven days) after initial hemorrhage, often when patient appears to be improving from initial symptoms associated with hemorrhage

d. **Hydrocephalus** — blockage of cerebrospinal fluid flow in subarachnoid space acutely due to blood or after recovery due to fibrosis

C. **Atherosclerotic fusiform aneurysm** — tortuous enlargement of intracranial vessels (particularly basilar artery or internal carotid artery) resulting in compression of adjacent structures

D. Post-traumatic carotid-cavernous sinus aneurysm (fistula)

1. Traumatic injury to wall of intracavernous portion of internal carotid artery allowing communication with cavernous sinus

   2. Produces massive shunting of blood into venous system with intracranial vascular congestion, unilateral pulsating exophthalmos, edema of eyelids, and audible murmur over orbit

E. Dissecting aneurysm of intracranial arteries

   1. Sequelae of blunt neck or head injury or due to atherosclerosis with intimal disruption (tear) allowing blood to dissect under and elevate intima

   2. Results in vascular occlusion with consequent brain ischemia and infarction

XI. **Arteriovenous malformations**

   1. **Developmental abnormality** consisting of **tangle of dilated blood vessels** forming abnormal communication between arterial and venous systems; many **vessels are thin-walled** and prone to rupture

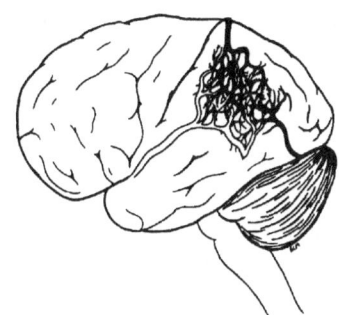

Cerebral arteriovenous malformation with feeding artery and large draining veins.

   2. Clinical symptoms — **seizures, recurrent headaches** (mimicking migraine), or **hemorrhage** (parenchymal hemorrhage or subarachnoid hemorrhage)

   3. Pathologic features — tangle of large arteries, arterioles, and arteriolized veins (many with marked intimal hyperplasia) embedded in gliotic brain parechyma

XII. **Subependymal and intraventricular hemorrhage** in premature infants

   1. Hemorrhage originates in highly vascular subependymal germinal matrix which is present until 35 weeks gestational age

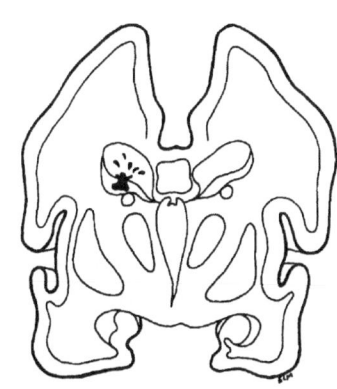

   2. Hypoxia, ischemia, and hypotension injure vascular endothelium resulting in hemorrhage

Subependymal germinal matrix hemorrhage can rupture into lateral ventricle.

60

SUGGESTED ADDITIONAL READING

Ausman JI, Shrontz CE, Pearce JE, Diaz FG, Crecelius JL: Vertebrobasilar insufficiency. A review. *Arch Neurol* 1985; 42:803-808.

Brey RL, Hart RG, Sherman DG, Tegeler CH: Antiphospholipid antibodies and cerebral ischemia in young people. *Neurology* 1990; 40:1190-1196.

Challa VR, Moody DM, Bell MA: The Charcôt-Bouchard aneurysm controversy: impact of a new histologic technique. *J Neuropathol Exp Neurol* 1992; 51:264-271.

Chuaqui R, Tapia J: Histologic assessment of the age of recent brain infarcts in man. *J Neuropathol Exp Neurol* 1993; 52:481-489.

Ferbert A, Brückmann H, Drummen R: Clinical features of proven basilar artery occlusion. *Stroke* 1990; 21:1135-1142.

Garcia JH: The evolution of brain infarcts. A review. *J Neuropathol Exp Neurol* 1992; 51:387-393.

Hart RG, Kanter MC: Hematologic disorders and ischemic stroke. *Stroke* 1990; 21:1111-1121.

Raps EC, Rogers JD, Galetta SL, Solomon RA, Lennihan L, Klebanoff LM, Fink ME: The clinical spectrum of unruptured intracranial aneurysms. *Arch Neurol* 1993; 50:265-268.

Ross R, Glomset JA: The pathogenesis of atherosclerosis. *N Engl J Med* 1976; 295:369-376 & 420-425.

Salgado AV, Furlan AJ, Keys TF: Mycotic aneurysm, subarachnoid hemorrhage, and indications for cerebral angiography in infective endocarditis. *Stroke* 1987; 18:1057-1060.

Toole JF: *Cerebrovascular Disorders*, 4th edition, New York, Raven Press, 1990.

Torvik Ak, Svindland A, Lindboe CF: Pathogenesis of carotid thrombosis. *Stroke* 1989;20:1477-1483.

CHAPTER 5: PATHOLOGY OF CNS INFECTIONS

I. Sites of central nervous system infections

    A. **Meningitis** — leptomeninges (pia and arachnoid) and subarachnoid space

    B. **Brain abscess** — localized within brain parenchyma

    C. **Subdural empyema** — within subdural space

    D. **Epidural empyema** — within epidural space

    E. **Ventriculitis**, ependymitis, choroid plexitis — within ventricular cavity or lining structures

    F. **Encephalitis** — diffuse involvement of brain parenchyma

II. **Blood-brain barrier**

    A. Limits entry of materials (such as infectious microorganisms) from blood into brain

    B. Brain capillaries are basis of blood-brain barrier — capillaries characterized by absence of fenestrations, minimal endothelial pinocytosis, and tight junctions between endothelial cells

III. **Meningitis**

    A. Cerebrospinal fluid findings — increased numbers of cells, along with variable changes in protein or glucose levels

        1. **Bacterial infection** — cell counts above 1000 cells/μL, composed mostly of **neutrophils**; increased protein level and **low glucose level**

2. **Viral infection** — slight increase in cell count, composed mostly of **lymphocytes**; normal protein level and **normal glucose level**

B. Unique features

1. Cerebrospinal fluid cell counts less than 1000 cells/μL indicate poor prognosis in *Haemophilus influenzae* and pneumococcal meningitis

2. Complications characteristic of particular organism

   a. **Meningococcemia** (*Neisseria meningitidis*) — purpuric rash, septicemia, shock and death associated with **adrenal hemorrhage**

   b. Pneumococcal meningitis (*Streptococcus pneumoniae*) — high mortality rate in alcoholics or in presence of systemic diseases

C. Etiologic agents by age group

1. Neonatal period — group B streptococcus, *Escherichia coli*, *Listeria monocytogenes*

2. 1-10 years — *Haemophilus influenzae, Neisseria meningitidis, Streptococcus pneumoniae*

3. 11-20 years — *Neisseria meningitidis, Streptococcus pneumoniae*

4. 30-60 years — *Streptococcus pneumoniae*

5. >60 years — gram negative organisms, *Streptococcus pneumoniae*

D. Pathologic features

1. Purulent exudate in subarachnoid space composed mostly of neutrophils, with increasing number of lymphocytes and monocytes as disease progresses

2. Inflammatory cells can insinuate into perivascular Virchow-Robin spaces

3. Inflammatory infiltration of blood vessel walls (vasculitis) can result in thrombosis with ischemic infarction in distribution of that vessel

E.   Specific conditions

1.   *Haemophilus influenzae* **meningitis**

   a.   Infection by small **gram negative coccobacillus** (*Haemophilus influenzae*), that requires factor V (nicotinamide dinucleotide) and factor X (hematin)

   b.   Organism can produce surrounding **polysaccharide capsule**; six capsular types, of which **type b** causes most infections; non-encapsulated organism commonly colonizes nasopharynx

   c.   During infancy, immune system normally has reduced ability to mount immunologic response against polysaccharide capsule; thus, most infections occur in infants and young children; recently introduced, genetically-engineered vaccine against *Haemophilus influenzae* type b can intensify immunologic response and produce immunity even in infants

2.   **Meningococcal meningitis**

   a.   Infection by anaerobic **gram-negative diplococcus** (*Neisseria meningitidis*); organism can produce **polysaccharide capsule**

   b.   Highly infectious organism spread in epidemics by respiratory route

   c.   Can present as either meningitis or meningococcemia (purpuric skin rash, shock, adrenal hemorrhage)

3.   **Pneumococcal meningitis**

   a.   Infection by **gram-positive diplococcus** (*Streptococcus pneumoniae*); organism can produce **polysaccharide capsule**

   b.   Part of normal upper respiratory tract flora; infectious foci in lungs, middle ear, or sinuses are common antecedents to meningitis; pneumococcal meningitis often follows basilar skull fracture which results in fistulous communication between subarachnoid space and sinuses or ear

   c.   Increased risk of systemic pneumococcal infection in individuals following surgical splenectomy or in sickle cell disease ("autosplenectomy" — sickling of red blood cells in spleen produces multiple splenic infarcts with ultimate destruction of spleen)

4. **Listeria meningitis**

   a. Infection by **aerobic gram-positive coccobacillus** (*Listeria monocytogenes*)

   b. Organism is ubiquitous in soil, dust, and water; produces infection mainly in **neonates** or in debilitated or immunocompromised patients

IV. **Focal suppurative infections**

A. **Brain abscess**

   1. Focal suppurative (purulent) **intraparenchymal infection,** usually secondary to seeding of brain parenchyma by organisms from chronic ear, sinus, pulmonary, or cardiac valvular infections

   2. Frequently produced by **anaerobic bacteria** — *Bacteroides* or *Peptostreptococcus*

   3. Presentation as intracranial **mass lesion,** hemiparesis, seizures

   4. Pathologic features

      a. **Cerebritis** — poorly defined focus (usually found in white matter or at gray matter-white matter junction) of soft necrotic (liquefying) brain tissue infiltrated by neutrophils and containing rapidly multiplying organisms

      b. **Abscess cavity** — by 7 days, margin of proliferating capillaries, collagen deposition, and astrocytic hypertrophy begins to demarcate abscess cavity ("ring-enhancing" lesion often visible by computed tomographic scans or magnetic resonance imaging)

      c. **Encapsulation** — by 2 to 4 weeks, fibroblastic proliferation around margin of abscess cavity produces encapsulation; greatest thickness of abscess capsule is toward pial surface (subarachnoid space) and thinnest portion is closest to ventricle (this is basis for fatal rupture of abscess into ventricle in some cases)

B. **Subdural empyema**

    1. Focal suppurative infection in subdural space, spreading over convexity, but limited to one side by falx cerebri

    2. Usually results from spread of infection from adjacent sinusitis, otitis, or skull osteomyelitis or following skull fracture

    3. Poor response to antibiotics due to poor accessibility of this avascular space

C. **Ventriculitis** — suppurative infection of ventricular cavities that can complicate meningitis, shunted hydrocephalus, or result from trauma

V. **Fungal and parasitic central nervous system infections**

A. **Toxoplasmosis**

    1. Infection by **protozoan organism** *Toxoplasma gondii* which is obligatory intracellular parasite; cats are definitive host

    2. Infection acquired by consuming undercooked meat contaminated with cysts which contain hundreds of organisms

    3. Primary infection in immunocompetent individuals often is asymptomatic or results in **transient malaise and lymphadenopathy**

    4. **Congenital toxoplasmosis**

        a. Primary infection in pregnant women results in **transplacental infection of fetus**

        b. Clinical features include **chorioretinitis, hepatosplenomegaly,** jaundice, multifocal brain necrosis (particularly periventricular) with subsequent mineralization and hydrocephalus

        c. Pathologic features — lymphocytic infiltration of leptomeninges, foci of necrosis and mineralization, glial nodules, and free organisms (tachyzoites) and parasite-laden cysts

    5. Infection in immunocompromised patients such as those with acquired immunodeficiency syndrome (AIDS) — multifocal necrotic lesions throughout brain, often containing abundant organisms (both free organisms and cysts)

B.   **Candidiasis**

   1.   Infection by opportunistic **fungus** from genus *Candida*; **most common fungal infection of brain**; usually due to species *Candida albicans*

   2.   **Disseminated** infections develop in immunosuppressed patients or patients with **lymphomas** or **diabetes**

   3.   Organism **initially infects skin or mucous membranes** (including gastrointestinal tract or vulva), then invades small superficial blood vessels to become hematogenously disseminated; **fungal emboli produce brain microinfarcts** which then serve as sites for development of fungal microabscesses containing both yeast forms and pseudohyphae

      a.   **Yeast** — single round to oval fungal cell (cytoplasm surrounded by cell wall) which can contain buds of daughter organisms

      b.   **Pseudohyphae** — individual fungal cells attached in strands producing irregular, bead-like appearance (constrictions between individual fungal cells help to distinguish pseudohyphae from smooth true hyphae)

C.   **Cryptococcosis**

   1.   Infection by opportunistic budding **yeast** *Cryptococcus neoformans* (formerly called *Torula histolytica* or European blastomycosis) found in soil contaminated by bird excreta

   2.   Primary pulmonary infection is followed (in susceptible or immunocompromised individuals) by **chronic basilar meningitis**

   3.   **Polysaccharide capsule** of organism impedes inflammatory response; organism visible in cerebrospinal fluid **"India ink preparation"** in which large unstained capsule produces halo around organism

   4.   Distention of perivascular Virchow-Robin spaces by mucoid deposits of polysaccharide capsular material produces honeycomb cystic appearance of brain parenchyma

   5.   Fibrosis in basilar subarachnoid cisterns results in **communicating hydrocephalus**

D. **Aspergillosis**

    1.   Infection by opportunistic saprophytic **fungus** from genus *Aspergillus*; characterized by septate hyphae which branch at acute angles

    2.   Organism is **vasoinvasive**; in immunocompromised individuals, **primary pulmonary infection** is followed by hematogenous dissemination; **fungal emboli produce brain microinfarcts** which then serve as sites for development of **fungal microabscesses** containing large numbers of hyphae

    3.   Clinical symptoms — focal neurologic deficits from infarcts secondary to occlusion of brain arteries by fungal emboli

VI. **Neurosyphilis**

    A.   Infection by **spirochete** *Treponema pallidum* transmitted by **venereal contact**, blood transfusion, or **transplacental transfer** from infected mother

    B.   Neurologic disease occurs late in patients with inadequately treated syphilis (**tertiary syphilis**)

    C.   Diagnosis based upon examination of cerebrospinal fluid

        1.   Nonspecific increases in cerebrospinal fluid cell count (lymphocytes and plasma cells), total protein level, and γ-globulin level; glucose levels are usually normal

        2.   Serologic tests

            a.   Non-treponemal (reagin) antibody test — Venereal Disease Research Laboratory (**VDRL**) slide test

            b.   Specific treponemal antibody tests — fluorescent treponemal antibody absorption (**FTA-ABS**) test or *Treponema pallidum* immobilization (TPI) test

    D.   Clinical syndromes of **neurosyphilis**

        1.   Asymptomatic neurosyphilis

            a.   No neurologic signs and symptoms; diagnosis based upon positive cerebrospinal fluid findings

b.   Pathologic features — lymphocytic infiltration of leptomeninges and non-caseating granulomas

2.   **Meningovascular syphilis**

a.   **Strokes** due to **arteritis** (spinal infarction can also occur); occurs 5-10 years after initial infection

b.   Pathologic features

(1)   Meningeal infiltration by mononuclear inflammatory cells

(2)   **Heubner's arteritis** (endarteritis) — lymphoplasmacytic infiltration around arteries, exuberant fibrous intimal proliferation, and thinning of media (with preservation of elastic lamina); leads to multiple ischemic infarcts in brain

3.   **General paresis** (general paresis of the insane; dementia paralytica)

a.   Progressive behavioral and **personality disturbances, dementia**, dysarthria, myoclonic jerks, seizures, spasticity, and **Argyll Robertson pupil**

(1)   Occurs 10-15 years after initial infection; associated with chronic meningitis and hydrocephalus

(2)   **Argyll Robertson pupil** — pupils bilaterally are small and irregular, do not dilate to mydriatic drugs or constrict (react) to light, but constrict on accommodation ("accommodates, but doesn't react")

b.   Pathologic features

(1)   Chronic meningoencephalitis with cerebral atrophy, marked generalized loss of neurons, and astrocytosis

(2)   **Rod cells** — proliferation of elongated microglial cells with nucleus often arranged perpendicularly to pia

(3)   **Gumma** — rare finding of masses of proliferating fibroblasts, lymphocytes, plasma cells, and multinucleated giant cells, acting as space occupying lesion

4.   **Tabes dorsalis**

    a.   Progressive **sensory loss** involving proprioception (dorsal column sensation) with resultant **ataxia** (sensory ataxia); **positive Romberg test** (falling from standing position after eye closure)

    b.   Lesser degree of loss of pain and temperature sense

        (1)   **Perforating ulcers** (particularly in feet)

        (2)   **Charcot joints — joint destruction** from repeated injury to relatively anesthetic joints

        (3)   **Lightning pains** — brief sharp stabbing (lancinating) pains, most frequent in legs

    c.   Usually occurs 15-20 years after initial infection

    d.   Pathologic features — **destruction (with minimal inflammation) and fibrosis of dorsal spinal roots** (particularly thoracic and lumbar) with **degeneration of spinal cord posterior columns** resulting in atrophic and dorsally-flattened spinal cord; **dorsal root ganglia are not involved**

VII.   Lyme disease

A.   Infection by **spirochete** *Borrelia burgdorferi* transmitted by **bite of ixodid tick** (deer tick); initial symptoms of flu-like illness (sometimes with **erythema chronicum migrans skin rash**)

B.   Clinical symptoms — **chronic meningitis** and **radiculitis** can present as **facial palsy**, other cranial nerve palsies, or spinal root syndromes; **chronic infection** produces **encephalitis** and **demyelination** that can mimic multiple sclerosis or neurodegenerative dementia

C.   Pathologic features — perivascular mononuclear inflammatory cell infiltration of peripheral and central nervous system with axonal degeneration and microglial nodules

VIII.   Acquired immunodeficiency syndrome (AIDS)

A.   Infection by **retrovirus human immunodeficiency virus-1 (HIV-1)** transmitted by venereal contact, blood transfusion, or transplacental transfer from infected mother

1. **RNA virus** which encodes for enzyme **reverse transcriptase (RNA-dependent DNA polymerase)** that synthesizes of complimentary DNA intermediate which acts as provirus, persisting within host cell and integrating into host cell DNA

2. Proviral DNA can remain dormant or transcribe itself into viral RNA, which directs protein synthesis to form new viral particles that bud from cell surface; common viral structural protein groups include:

   a. *gag* — core antigens p25, p9, p17

   b. *pol* — reverse transcriptase, endonuclease, p66, p51

   c. *env* — major envelope glycoproteins gp120, gp41

B. Clinical symptoms

   1. Initial infection — transient malaise, lymphadenopathy, and occasionally aseptic meningitis

   2. Seropositivity (serum antibody response to organism) develops within several months of initial infection

   3. Clinical latency — clinically asymptomatic carrier state which can last for months to years (sometimes 10 or more years)

      a. HIV-1 organism binds specifically to T4 (CD4) receptor on peripheral thymus-derived "helper" lymphocytes, which allows organism to enter this cell and eventually kill it

      b. **Altered T4:T8 (helper:suppressor) ratio — progressive depletion** of peripheral T4 "helper" lymphocytes results in altered ratio of T4 lymphocytes to T8 (CD8) "suppressor" lymphocytes

   4. **AIDS** — clinical illness develops when T4:T8 ratio is sufficiently low that opportunistic infections and neoplasms develop

C. **Neuropathologic features of AIDS**

   1. **HIV encephalitis** — multiple disseminated foci of **microglial nodules** and **multinucleated giant cells** (derived from HIV-infected blood macrophages), particularly in cerebral and cerebellar white matter, basal ganglia, and thalamus; associated with **diffuse white matter**

pallor and reactive astrocytosis; clinical correlate is **progressive dementia (AIDS-dementia complex)**

2.   **Vacuolar myelopathy** — spongy vacuolation of myelin in spinal cord dorsal and lateral columns

3.   **Peripheral neuropathy** — axonal degeneration, segmental demyelination, and mononuclear inflammatory cell infiltrates

4.   **Myopathy**

   a.   Inflammatory myopathy, type II muscle fiber atrophy, neurogenic atrophy (denervation)

   b.   **AZT myopathy — mitochondrial abnormalities** (including mitochondrial proliferation, abnormal mitochondrial shapes, intramitochondrial paracrystalline inclusions), **secondary to treatment with azidothymidine** (AZT; zidovudine)

5.   **Opportunistic infections**

   a.   Parasitic infections (**toxoplasmosis**), fungal infections (*Cryptococcus neoformans, Candida* species, *Aspergillus* species), and spirochetal infection (*Treponema pallidum*)

   b.   **Cytomegalovirus** (CMV) infection

      (1)   Infection by ubiquitous double-stranded DNA virus of herpesvirus group (**cytomegalovirus)**

      (2)   Infects **vascular endothelial cells** resulting in **vasculitis** with ischemic infarcts in brain, peripheral nerves, and spinal nerve roots (radiculitis)

      (3)   Microglial nodules in brain associated with cells (astrocytes or neurons) containing **viral inclusions**

         (a)   **Cowdry type A intranuclear inclusion** — ovoid dense eosinophilic intranuclear bodies surrounded by clear halo (separating inclusion from nuclear membrane)

         (b)   **Cytomegalic cell** — ballooned cell with granular basophilic cytoplasm

c.   **Progressive multifocal leukoencephalopathy (PML)**

   (1)   Infection by opportunistic small double-stranded DNA virus of **papovavirus** group B (polyomaviruses: JC virus and SV40 virus)

   (2)   Clinical symptoms — rapidly evolving dementia, ataxia, visual field defects, spasticity and weakness, **swallowing and speech difficulties**, coma, and **death usually within 6 months** of onset

   (3)   Pathologic features

      (a)   Multiple large areas of discoloration and granular (crumbling) destruction of white matter

      (b)   Microscopic evidence of white matter destruction with little inflammatory response; **numerous giant astrocytes with bizarre-shaped nuclei** and enlarged oligodendrocytes with basophilic glassy inclusions

6.   **Lymphomas** — perivascular, often multifocal B-cell lymphomas; histologic large cell or mixed large and small cell lymphoma

IX.   **Arbovirus encephalitis**

A.   Infection with single-stranded RNA viruses transmitted by blood-sucking arthropods, including mosquitoes, sandflies, and ticks

B.   Includes eastern equine encephalitis, western equine encephalitis, Venezuelan equine encephalitis, St. Louis encephalitis

C.   Pathologic features

   1.   Aseptic meningitis — mononuclear inflammatory cell infiltration of leptomeninges with perivascular "cuffing" by inflammatory cells (inflammatory cells in perivascular Virchow-Robin spaces)

   2.   Neuronophagia — degenerating neurons surrounded by mononuclear inflammatory cells

   3.   Microglial nodules — dense collections of microglial cells

X.   **Herpesvirus encephalitis**

A.   Reactivation of latent infection by double-stranded DNA virus **herpes simplex virus (HSV) type 1 or 2**

B.   During **primary infection of skin or mucous membranes**, viral particles are transported by retrograde axoplasmic transport in peripheral nerves to sensory ganglia where **persistent latent infection** is established

C.   **Reactivation of latent infection** usually produces painful, vesicular, ulcerated lesions on skin or mucosa which is target of involved nerve

D.   Encephalitis results from transport into brain of virus particles during reactivation (through nerve branches from trigeminal ganglia that innervate leptomeninges of middle and anterior cranial fossa)

E.   Clinical symptoms

1.   Prodrome of fever, headache, pharyngitis, and malaise is followed by rapidly developing personality and behavioral changes, seizures, and coma

2.   Bloody cerebrospinal fluid can be obtained by lumbar puncture

3.   Electroencephalogram (EEG) shows focal slow waves in temporal area

4.   Computed tomographic (CT) scan or magnetic resonance imaging (MRI) shows hemorrhagic destruction centered around anterior inferior temporal lobes

F.   Pathologic features — **hemorrhagic necrosis of brain** with dense mononuclear inflammatory cell infiltrate and **Cowdry type A intranuclear inclusions** in neurons, astrocytes, and oligodendrocytes

XI. **Poliomyelitis**

A.   Infection by **enterovirus** (small RNA virus of picornavirus group); enteroviruses include: poliovirus, echovirus, and coxsackievirus; most cases of poliomyelitis result from poliovirus infection

B.   Primary infection is gastrointestinal and in most cases is either asymptomatic or results in "minor illness" of pharyngitis, fever, nausea, vomiting, or diarrhea

C.   In some individuals, viremia can be followed by "major illness"

1.   Viral particles gain entrance to central nervous system through areas that normally lack blood-brain barrier (such as choroid plexus, area postrema) producing **aseptic meningitis**

2.   Virus enters and kills susceptible neurons (presumably through cell receptors), including anterior horn cells, cranial nerve motor neurons, Betz cells of primary motor cortex (precentral gyrus), resulting in paralysis; neurons of reticular formation, brain stem nuclei, thalamus, and hypothalamus can also be involved

3.   Skeletal muscle atrophy follows denervation

D.   Pathologic features — destruction of involved neurons, neuronophagia, microglial nodules, and perivascular mononuclear inflammatory cell infiltrates

XII.   **Rabies**

A.   Infection by single-stranded ("negative-stranded") RNA virus; virus possesses RNA-dependent RNA transcriptase which synthesizes "positive" RNA strand that directs protein synthesis and viral replication

B.   Virus is transmitted to humans by **bites from infected animals** with incubation period varying from one to three months; virus is transferred to central nervous system by **retrograde axoplasmic transport** in peripheral nerve motor and sensory axons terminating near area of bite; incubation periods are longer in leg bites and shorter in facial bites

C.   Clinical symptoms — subacute onset (over 3 to 4 days) of **personality change, hyperexcitability,** and **headache** followed by **dysphagia** and **pharyngeal muscle spasm** (leading to drooling or "frothing at the mouth" and inability to swallow water or "hydrophobia"), **facial muscle spasms, seizures,** and **coma**; nearly **universally fatal** after one to two weeks

D.   Pathologic features

1.   Accumulation of viral RNA in cytoplasm of **infected neurons** produces **cytoplasmic inclusions**

a.   **Negri bodies** — ovoid eosinophilic hyaline **cytoplasmic bodies** which have small internal basophilic granules; particularly prominent in large neurons such as hippocampal pyramidal cells, cerebellar Purkinje cells, and neurons of brain stem cranial nerve nuclei

        b.   **Lyssa bodies** — irregular or ovoid homogeneous cytoplasmic bodies; similar to Negri bodies but without internal structure and often present in smaller neurons

    2.   Perivascular mononuclear inflammatory cell infiltrates, neuronal degeneration, and neuronophagia; however, inflammation is minimal in areas containing large numbers of Negri bodies or lyssa bodies

XIII.   Subacute sclerosing panencephalitis (SSPE)

    A.   Infection by single stranded ("negative-stranded") RNA virus of paramyxovirus (parainfluenza virus) group; results from **mutant measles virus** lacking gene coding for M-protein (mutation prevents viral assembly and release and allows intracellular persistence of virus particles)

    B.   Clinical symptoms

        1.   Slowly progressive degenerative disorder developing several years after initial measles infection and associated with markedly **elevated measles antibody titers** in cerebrospinal fluid

        2.   Initial personality changes and **intellectual deterioration** progress to **myoclonic seizures, ataxia, blindness**, and death after several months or years

    C.   Pathologic features

        1.   Extensive **loss of myelin and axons in cerebral white matter**, particularly in deeper white matter with relative sparing of U-fibers

        2.   Minimal perivascular cuffs of mononuclear inflammatory cells

        3.   Marked astrocytic proliferation and hypertrophy in destroyed areas of white matter

        4.   Prominent **Cowdry type A intranuclear inclusions** in oligodendrocytes, neurons, and occasionally astrocytes; eosinophilic cytoplasmic inclusions also frequently present

XIV. Transmissible spongiform encephalopathies

    A. Group of human and animal diseases resulting from **prions** and transmitted through ingestion or inoculation of infected brain tissue; includes animal diseases such as scrapie, transmissible mink encephalopathy, and bovine spongiform encephalopathy ("mad cow disease") and human diseases of Kuru, **Creutzfeldt-Jakob disease**, Gerstmann-Sträussler-Scheinker syndrome, and fatal familial insomnia

    B. Clinical symptoms — **rapidly progressive dementia** and ataxia and myoclonus in some cases; electroencephalogram (EEG) shows high voltage slow-sharp wave complexes on nearly flat background (**"burst suppression" pattern**)

    C. Causative agent is **prion** (PrP) which is composed of **protein** only (no nucleic acids)

        1. In humans, chromosome 20 contains gene sequence for analog of prion protein which is present in normal cells (function unknown)

        2. Point mutations of prion protein gene lead to disease in familial cases of spongiform disorders

        3. Abnormal protein seems to be able to instruct normally formed protein to change into abnormal protein (post-translational alteration in protein structure)

        4. Abnormal prion protein is **resistant to standard sterilization procedures** (boiling, formalin, alcohol, or ultraviolet radiation), but inactivated by autoclaving or immersion in sodium hypochlorite (household bleach) or sodium hydroxide

        5. Pathologic features

            a. **Spongiform changes** in neuropil, marked neuronal loss, and gliosis; spongiform change consists of **numerous small vacuoles** throughout neuropil related to dilation of neuronal and astrocytic processes

            b. **Kuru plaques** — hyaline eosinophilic masses with radiating peripheral amyloid fibrils; composed of prion protein

            c. **Absent cellular or humoral inflammatory response**

77

SUGGESTED ADDITIONAL READING

Bell WE, McCormick WF: *Neurologic Infections in Children.* ed. 2. Philadelphia, W. B. Saunders Co, 1981.

Budka H: Neuropathology of human immunodeficiency virus infection. *Brain Pathology* 1991; 1:163-175.

Burns DK: The neuropathology of pediatric acquired immunodeficiency syndrome. *J Child Neurol* 1992; 7:332-346.

Hesselink JR (ed): Infectious and inflammatory diseases. *Neuroimag Clin NA* 1991; 1(1):1-257.

Johnson RT: *Viral Infections of the Nervous System.* New York, Raven Press, 1982.

Prusiner SB: Genetic and infectious prion diseases. *Arch Neurol* 1993; 50:1129-1153.

Vinken PJ, Bruyn GW, Klawans HL, Harris AA (eds): *Microbial Disease. Handbook of Clinical Neurology. Volume 52.* Amsterdam, Elsevier Science Publishers, 1988.

Vinken PJ, Bruyn GW, Klawans HL, McKendall RR (eds): *Viral Disease. Handbook of Clinical Neurology. Volume 56.* Amsterdam, Elsevier Science Publishers, 1989.

Wood M, Anderson M: *Neurological Infections.* London, W. B. Saunders Co, 1988.

CHAPTER 6: NEURODEGENERATIVE DISEASES AND DEMENTIA

I.   Definitions

    A.   **Neurodegenerative disease**

        1.   Category of disorders without known etiology

        2.   Following identification of etiology, some disorders previously considered neurodegenerative are now categorized as infectious (for example, Creutzfeldt-Jakob disease) or metabolic (for example, Wilson's disease)

    B.   **Dementia**

        1.   Descriptive term for **deterioration of intellectual/cognitive abilities of sufficient severity to interfere with normal social functioning**

        2.   Signs of dementia include **confusion, memory disturbances, difficulties with problem solving and abstract thinking, impaired judgement,** personality changes, and emotional lability

SELECTED NEURODEGENERATIONS
Alzheimer's disease
Ataxia telangiectasia
Diffuse Lewy body disease
Friedreich's ataxia
Huntington's disease
Parkinsonism-dementia-ALS of Guam
Parkinson's disease
Pick's disease
Progressive supranuclear palsy

        3.   Although dementia can occur at any age, term is often used to imply neurodegenerative disease affecting older individuals

II.   **Alzheimer's disease**

    A.   Most common neurodegenerative dementia; affects up to 10% of surviving population over age 65 years

B. Diagnosis

    1. *Probable* Alzheimer's disease — **clinical evidence of progressive dementia** with **no disturbance of consciousness** and **absence of systemic or other brain diseases** that cause dementia

    2. *Definite* Alzheimer's disease — neuropathologic (biopsy or autopsy) confirmation of clinical diagnosis

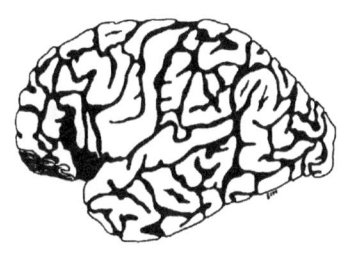

Generalized cerebral cortical gyral atrophy and marked sulcal enlargement in Alzheimer's disease.

C. Pathologic features

    1. Nonspecific generalized cerebral cortical atrophy and ventricular enlargement

    2. **Neurofibrillary tangles**

        a. Intracellular clumps of 10 nm filaments in double helical arrangement (**paired helical filaments**); occupy cerebral cortical neuronal cell body and axonal and dendritic processes

        b. **Argentophilic** staining properties (identified with silver staining techniques such as Bodian or Bielschowsky stains)

Neurofibrillary tangle in neuronal cell body and extracellular neuritic plaque associated with dendritic processes.

        c. Neurofibrillary tangles can also be found in other disorders:

            (1) **Dementia pugilistica** ("punch drunk syndrome") — in cerebral cortex, midbrain substantia nigra, and pontine locus ceruleus

            (2) **Progressive supranuclear palsy** — in basal ganglia, brain stem, and cerebellum

(3) **Parkinsonism-dementia-amyotrophic lateral sclerosis syndrome of Guam** — in cerebral cortex

3. **Neuritic (senile) plaques**

a. **Core** of extracellular **amyloid** (ß/A4-amyloid) surrounded by abnormal ("dystrophic") neurites (nerve cell processes containing distended lysosomes and paired helical filaments), microglia (brain macrophages), and reactive astrocytes

Neurofibrillary tangle composed of intracellular paired helical filaments.

b. Found in cerebral cortex (particularly hippocampus and amygdala)

c. Identified with silver stains (Bodian or Bielschowsky stains) or with amyloid stains (Congo red or thioflavin)

d. Specific for Alzheimer's disease

4. **Generalized cerebral cortical and subcortical nerve cell loss**, particularly marked in hippocampus and amygdala; loss in specific nuclei explains some clinical symptoms:

a. Memory disturbance — loss of cholinergic neurons of nucleus basalis of Meynert (which send processes to hippocampus)

Neuritic plaque with extracellular amyloid core.

b. Emotional (depressive) symptoms — loss of noradrenergic neurons of pontine locus ceruleus and serotonergic neurons of pontine raphe nuclei (which send processes to cerebral cortex)

5. Granulovacuolar degeneration of Simcowicz

a. Numerous small basophilic granules surrounded by clear halo in hippocampal pyramidal neurons

b. Can also be found in Pick's disease and parkinsonism-dementia-amyotrophic lateral sclerosis syndrome of Guam

6. Hirano bodies

    a. Intracellular, short, eosinophilic rod-like structures in hippocampal pyramidal neurons

    b. Can also be found in Pick's disease and parkinsonism-dementia-amyotrophic lateral sclerosis syndrome of Guam

D. Possible genetic factors

    1. **Autosomal dominant inheritance pattern in some families** with linkage (in different families) to genes on chromosome 21, chromosome 19, or chromosome 14

    2. **Clinical dementia** with **histopathologic findings of Alzheimer's disease** is found in virtually all patients with **Down's syndrome** (trisomy 21) living beyond age 30 years

E. No definitive treatment is currently available, but supportive care for patient and family can be provided

III. **Pick's disease** (lobar atrophy)

A. Clinical symptoms — **progressive dementia** with early **prominent alterations in emotion, affect, and personality**, but often clinically indistinguishable from Alzheimer's disease

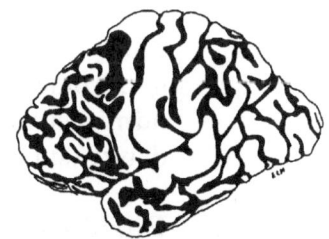

Pick's disease with marked frontotemporal gyral atrophy and sulcal enlargement.

B. Pathologic features

    1. **Relatively localized severe frontotemporal cerebral gyral atrophy** (hence, "lobar atrophy")

        a. **"Knife edge atrophy"** — extremely thin gyri due to severe gyral atrophy

        b. Relative **sparing of posterior two-thirds of superior temporal gyrus** (portion of superior temporal gyrus posterior to central sulcus)

    2. Marked loss of neurons and cortical and subcortical gliosis, particularly in frontotemporal areas; characteristic **loss of granular neurons of hippocampal dentate fascia**

3. **Pick bodies**

    a.   **Intraneuronal inclusion** in **ballooned neurons**; inclusion has argentophilic staining properties (in Bodian or Bielschowsky silver stains)

    b.   Most common in temporal lobe cortical pyramidal neurons and hippocampal dentate fascia granular neurons

    c.   Composed of mixture of accumulated neurofilaments, neurotubules, and endoplasmic reticulum

C.   Familial clustering of some cases suggestive of autosomal dominant inheritance

IV. Parkinsonism (Parkinson's syndrome)

A.   Clinical symptoms — **triad** of **resting tremor**, **rigidity**, and **bradykinesia**, along with postural, gait, and handwriting disturbances

SELECTED CAUSES OF PARKINSONISM
Diffuse Lewy body disease
Drug-induced (neuroleptics)
Multisystem atrophy
Parkinsonism-dementia-ALS of Guam
Parkinson's disease
Progressive supranuclear palsy
Toxin-induced (manganese; MPTP)
Traumatic (dementia pugilistica)
Wilson's disease

B.   Most common cause of Parkinson's syndrome is pharmacological due to administration of **dopamine antagonist** drugs (such as neuroleptics)

C.   **Parkinson's disease**

    1.   Idiopathic progressive neurodegenerative disorder

    2.   Pathologic features

        a.   Gross depigmentation of **midbrain substantia nigra** and **pontine locus ceruleus** due to loss of pigmented (melanin-containing) neurons

            (1)   Substantia nigra and locus ceruleus are normally unpigmented at birth

            (2)   Catecholaminergic neurons normally accumulate neuromelanin pigment resulting in gross pigmentation of these structures by early adolescence

b. Microscopic loss of pigmented neurons evidenced by collections of melanin-containing macrophages ("neuronal tombstones")

c. **Lewy bodies**

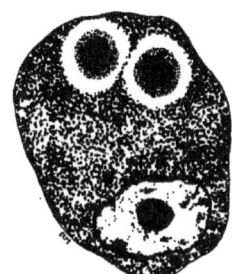

Pigmented neuron containing two Lewy bodies.

(1) Cytoplasmic inclusions in remaining neurons in substantia nigra and locus ceruleus

(2) Central dense hyaline eosinophilic core surrounded by halo of lighter amorphous eosinophilic material; often accompanied by diffuse amorphous eosinophilic areas (hyaline inclusions) that displace cytoplasmic melanin granules

(3) Composed of densely packed filaments in core and looser filamentous accumulation in peripheral halo

d. Loss of dopaminergic nerve fibers in striatum (basal ganglia) accompanied by >80% reduction in basal ganglia dopamine concentration

3. Pharmacologic treatment involves dopamine receptor agonist therapy

a. Administered **L-dopa** is converted to dopamine by surviving neurons resulting in increased striatal dopamine concentrations

b. Agonists **bromocriptine** or pergolide directly stimulate dopamine receptors

4. Dementia — occasional patients developing dementia have either diffuse Lewy body disease (Lewy body dementia) or concurrent Alzheimer's disease

D. **Multisystem atrophy** — group of degenerative diseases characterized by neuronal loss from various systems, with clinical and pathologic features in varying combinations (sometimes classified as "parkinsonism plus"); some cases have metabolic abnormalities of mitochondria (mitochondrial encephalomyopathies)

1. **Striatonigral degeneration**

   a. Clinical symptoms — parkinsonism unresponsive to dopaminergic therapy

   b. Pathologic features

      (1) Marked gross shrinkage and brown discoloration of putamen and loss of pigmentation of substantia nigra

      (2) Marked neuronal loss (and reactive gliosis) in putamen, caudate, and globus pallidus, with lesser neuronal loss in red nucleus and subthalamic nucleus

2. **Olivopontocerebellar atrophy** (OPCA)

   a. Clinical symptoms — progressive severe ataxia, external ophthalmoplegia, and facial palsy

   b. Pathologic features

      (1) Marked gross shrinkage of basis pons and middle cerebellar peduncle, atrophy of cerebellar cortex and white matter, and loss of olivary bulge in medulla resulting in unusual prominence of medullary pyramids

      (2) Severe neuronal loss in inferior olivary nuclei, marked loss of cerebellar Purkinje cells and granule cells (except relative sparing of midline cerebellum); dentate nuclei gliotic from loss of incoming Purkinje cell axons; normal superior cerebellar peduncle

3. **Shy-Drager syndrome** (idiopathic orthostatic hypotension)

   a. Clinical symptoms — progressive autonomic dysfunction with postural hypotension, bladder and bowel incontinence, loss of sweating, and impotence

   b. Pathologic features — loss of preganglionic sympathetic neurons from spinal cord intermediolateral cell column (lateral horn) and depigmentation of midbrain substantia nigra and pontine locus ceruleus

E. **Steele-Richardson-Olszewski progressive supranuclear palsy**

    1.   Clinical symptoms

        a.   **Parkinsonism unresponsive to dopaminergic therapy**

        b.   Progressive paralysis of vertical eye movements (**paralysis of downgaze**) and **pseudobulbar palsy**

        c.   **Subcortical dementia** — marked **apathy, inertia,** and **slowing of thought processes** with only minimal memory loss or altered judgement

    2.   Pathologic features

        a.   Gross **atrophy of superior colliculi and subthalamic nuclei** with depigmentation of substantia nigra

        b.   Marked neuronal loss (and reactive gliosis) in superior colliculi, subthalamic nuclei, and substantia nigra

        c.   **Neurofibrillary tangles in basal ganglia, thalamus, and brain stem and cerebellar gray matter nuclei;** no neurofibrillary tangles in cerebral cortex

F. **Diffuse Lewy body disease** (Lewy body dementia)

    1.   Clinical symptoms — **progressive dementia and parkinsonism**

    2.   Pathologic features

        a.   Gross **depigmentation of substantia nigra and locus ceruleus**

        b.   Neuronal loss with **Lewy bodies in remnant neurons in substantia nigra and locus ceruleus,** along with **Lewy bodies in cerebral cortical neurons**

G. **Parkinsonism-dementia-ALS syndrome of Guam**

    1.   Clinical symptoms — **progressive dementia, parkinsonism, and motor neuron disease** (upper motor neuron spasticity and lower motor neuron flaccid weakness) in various combinations

2.   Pathologic features

    a.   Neuronal loss in cerebral cortex, substantia nigra, locus ceruleus, basal ganglia, thalamus, and spinal cord anterior horns

    b.   Numerous **neurofibrillary tangles in cerebral cortex and brain stem**; no neuritic plaques

3.   Related to **ingestion of neurotoxin ß-N-methylamino-alanine from cycad beans** in Guamanian diet

H.   Experimental MPTP-induced parkinsonism

1.   **MPTP** (1-methyl-4-phenyl-1,2,3,6-tetrahydropyridine) — opiate analog "designer drug"

2.   **Astrocyte monoamine oxidase-B (MAO-B) enzymatically oxidizes MPTP to MPP$^+$** which is transported into dopaminergic neurons by dopamine uptake system

3.   **MPP$^+$ inhibits mitochondrial electron transport chain resulting in neuronal death**

4.   Resultant depigmentation of substantia nigra and marked loss of substantia nigra neurons without Lewy bodies

V.   **Huntington's disease (Huntington's chorea)**

A.   Clinical symptoms

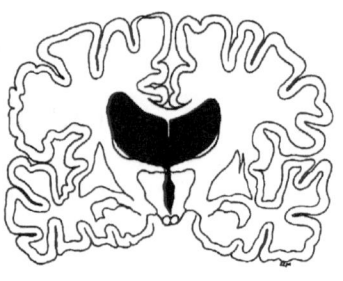

1.   **Progressive choreoathetosis and dementia**

2.   Psychiatric symptoms and suicide are common during early phase of disease

3.   Average onset is age 35 years

Marked caudate atrophy and ventricular enlargment in Huntington's disease.

4.   **Juvenile Huntington's disease (Westphal variant)** — onset before age 15 years; characterized by **rigidity and seizures**

B.   **Dominantly-inherited disorder** due to abnormal gene on short arm of chromosome 4; genetic abnormality consists of expansion (increased length) of unstable trinucleotide (CAG) repeat sequence within gene in region 4p16; increased length of trinucleotide repeat sequence correlates with earlier age of onset of disease

C. CT scan and MRI show **atrophy of caudate nucleus** with consequent ventricular enlargement ("box car ventricles")

D. Pathologic features

1. Marked gross **atrophy of caudate nucleus** and to lesser extent putamen and globus pallidus

2. Marked loss of **medium-sized striatal spiny neurons** which contain γ-aminobutyric acid (GABA), substance P, and enkephalin

3. Preservation of striatal aspiny neurons

VI. **Spinocerebellar degeneration** — heterogeneous group of heredofamilial disorders characterized by progressive **ataxia**

A. **Friedreich's ataxia**

1. Clinical symptoms

a. **Progressive limb and truncal ataxia** beginning before age 20 years

b. Dysarthria, **areflexia, spasticity** and extensor plantar response (Babinski reflex), **pes cavus foot deformity, kyphoscoliosis, and cardiomyopathy**

2. **Autosomal recessive inheritance** due to abnormal gene on chromosome 9

3. Pathologic features

a. Marked axonal degeneration in spinal cord posterior columns and ventral and dorsal spinocerebellar tracts

b. Degeneration of Clarke's column

c. Neuronal loss from dorsal root ganglia and degeneration of peripheral nerve large myelinated axons

B. **Ataxia-telangiectasia (Louis-Bar syndrome)**

  1. Clinical symptoms

  a. Progressive **ataxia** beginning in infancy with subsequent choreoathetosis, nystagmus, and areflexia

  b. Development of **telangiectasia** (blood vessel dilation) over conjunctiva, face, and neck

  c. Associated with **thymic and lymphoid hypoplasia, deficiency of serum IgA**, recurrent **respiratory infections**, and development of malignant **lymphoma**

  2. **Autosomal recessive inheritance**; genetically defective ability to repair damage to DNA produced by radiation (x-rays) or by normal intracellular free radical metabolites

  3. Pathologic features

  a. Severe axonal degeneration in spinal cord posterior columns and diffuse loss of Purkinje cells with axonal swellings (torpedoes) in remnant Purkinje cells

  b. **Large cells with large irregular bizarre nuclei** found in pituitary gland, dorsal root ganglia, and various visceral organs

SUGGESTED ADDITIONAL READINGS

Bastian FO (ed): *Creutzfeldt-Jakob Disease and Other Transmissible Spongiform Encephalopathies.* St. Louis, Mosby-Year Book, 1991.

Brumback RA, Leech RW: Alzheimer's disease: pathophysiology and the hope for therapy. *J Okla State Med Assoc* 1994; 87:103-111.

Cederbaum JM, Gancher ST (eds): Parkinson's disease. *Neurol Clin* 1992; 10(2):301-600.

Duckett S (ed): *The Pathology of the Aging Human Nervous System.* Philadelphia, Lea & Febiger, 1991.

Gilman S, Bloedel JR, Lechtenberg R: *Disorders of the Cerebellum.* Philadelphia, F. A. Davis Co, 1981.

Jankovic J, Tolosa E (eds): *Parkinson's Disease and Movement Disorders*. Baltimore, Urban & Schwarzenberg, 1988.

Katzman R, Rowe JW: *Principles of Geriatric Neurology*. Philadelphia, F. A. Davis Co, 1992.

Khachaturian ZS: Diagnosis of Alzheimer's disease. *Arch Neurol* 1985; 42:1097-1105.

Terry RD (ed): *Aging and the Brain*. New York, Raven Press, 1988.

Vinken PJ, Bruyn GW, Klawans HL (eds): *Extrapyramidal Disorders. Handbook of Clinical Neurology. Volume 49*. Amsterdam, Elsevier Science Publishers, 1986.

Vinken PJ, Bruyn GW, Klawans HW, Jong JMBU (eds): *Hereditary Neuropathies and Spinocerebellar Atrophies. Handbook of Clinical Neurology. Volume 60*. Amsterdam, Elsevier Science Publishers, 1991.

CHAPTER 7: NERVOUS SYSTEM TUMORS

I. Pathophysiology of symptoms

    A. Local effects of tumors

        1. **Infiltration, invasion,** and **destruction** of normal nervous system tissues by tumor — produces focal neurological signs

        2. Mass of tumor produces direct pressure on neural structures causing degeneration (although brain and spinal cord can adjust remarkably well to gradually increasing pressure)

        3. **Compromise of local circulation** (due to direct pressure on capillaries and small arteries and veins) can be associated with local tissue necrosis or more distant infarction

            a. In **spinal tumors,** symptoms can develop acutely when tumor **compression of anterior spinal artery** is sufficient to produce **ischemia and infarction of spinal cord**

                (1) **Anterior spinal artery syndrome** — spastic paraplegia and loss of pain and temperature sensation while posterior column proprioceptive sensation remains intact

        4. Brain **edema** — usually greatest around tumor; can interfere with functioning of more remote neural tissue, adding to clinical symptoms directly attributable to tumor mass and eventually (usually over period of months) resulting in demyelination and gliosis (astrocytic proliferation)

            a. Treatment with high doses of **corticosteroids** can reduce edema

            b. **Osmotic agents** (such as intravenous infusion of hypertonic mannitol solution) can very rapidly decrease intracranial pressure

5. **Seizures** — tumors involving cerebral cortex can result in focal or generalized seizures

    a. Oligodendrogliomas of temporal lobe are often associated with chronic seizures

6. Spinal tumor — symptoms depend on level of tumor involvement and whether tumor is extramedullary or intramedullary

PRINCIPAL BRAIN TUMORS BY LOCATION	
Cerebral hemisphere	**Fourth ventricle**
Astrocytoma	Choroid plexus papilloma
Ependymoma	Ependymoma
Meningioma	Meningioma
Metastatic carcinoma	**Pituitary region**
Oligodendroglioma	Craniopharyngioma
Vascular malformation	Pituitary adenoma
Optic chiasm	Meningioma
Astrocytoma	**Brain stem**
Meningioma	Astrocytoma
Pineal region	**Cerebellopontine angle**
Astrocytoma	Acoustic neuroma
Germ cell tumor	Epidermoid cyst
Pineoblastoma	Meningioma
Pineocytoma	**Spinal cord (epidural)**
Cerebellum	Metastatic carcinoma
Astrocytoma	**Spinal cord (extramedullary)**
Hemangioblastoma	Meningioma
Medulloblastoma	Neurilemoma
Metastatic carcinoma	Neurofibroma
Third ventricle	**Spinal cord (intramedullary)**
Choroid plexus papilloma	Astrocytoma
Colloid cyst	Ependymoma
Ependymoma	

   a. **Extramedullary tumors** — growth **outside spinal cord parenchyma** produces symptoms related to nerve root compression and bone destruction before spinal cord symptoms

   b. **Intramedullary tumors** — because of small size of spinal cord, **tumor growth within cord parenchyma** presents early as disturbance of spinal cord function (disruption of long tracts and local segmental signs)

B. General effects of tumors — from **increased intracranial pressure**

   1. Rigidity of cranium (following fontanelle and suture closure in childhood) allows no room for expansion; thus, increased pressure from tumor mass and surrounding brain edema is transmitted throughout ventricular system and brain

      a. **Obstructive hydrocephalus** — tumor mass and edema can distort ventricular system enough to **obstruct foramina** resulting in hydrocephalus; examples include:

         (1) Obstruction of foramen of Monro bilaterally by colloid cyst of third ventricle results in lateral ventricular enlargement

        (2)   Pressure on midbrain tectum by pineal region tumors obstructs aqueduct of Sylvius with resultant enlargement of lateral and third ventricules

    b.   **Herniation** — structures are displaced and distorted by edema and tumor mass resulting in protrusion (**herniation**) of portions of brain across midline (**transfalcial herniation** of cingulate gyrus), through tentorial incisura (**transtentorial herniation** of temporal lobe uncus), or through foramen magnum (**cerebellar tonsillar herniation**)

        (1)   Sudden exacerbation of signs and symptoms usually results from acute herniation or acute increased size of mass due to hemorrhage within tumor

        (2)   Death generally is from brain stem compression due to transtentorial herniation of medial temporal lobe or herniation of cerebellar tonsils through foramen magnum

  2.   **Papilledema** — increased cerebrospinal fluid pressure in subarachnoid space surrounding optic nerve produces **swelling of optic nerve head** (optic disc)

  3.   **Headache** — increased intracranial pressure produces bilateral headache that is increased by coughing or straining (Valsalva maneuver) and is greatest in morning or after lying down

II.   **Diagnosis** of brain tumor

  A.   Radiologic imaging studies (CT scan or MRI) — identifies tumor mass

  B.   **Cerebral angiography** — identifies vascular supply to tumor

  C.   **Lumbar puncture** — identifies **malignant cells** in cerebrospinal fluid in **meningeal carcinomatosis** or **leukemia**; lumbar puncture can result in **fatal herniation** by producing fluid shifts

III.   **Glioma**

  A.   **Astrocytoma**

    1.   **Diffusely infiltrative** tumor of **malignant astrocytes**; invades widely throughout brain, but does not metastasize outside central nervous system

a. Adults — **most common primary malignant brain tumor**; usually located in cerebral hemisphere, but can also occur as primary intramedullary spinal cord tumor

b. Children — located in brain stem, optic nerves, or cerebellum

2. Classification (grading) based on histologic appearance, including presence or absence of nuclear atypia, mitotic figures, endothelial hyperplasia, and necrosis:

Subpial and perineuronal accumulation of infiltrating malignant astrocytes.

a. Astrocytomas are **highly infiltrative tumors** with no distinct margin

(1) Perineuronal satellitosis — clustering of neoplastic astrocytes around neurons in gray matter

(2) Neoplastic astrocytes accumulate in subpial and subventricular areas; in high grade tumors, malignant astrocytes can rupture through ependymal or pial boundaries to be disseminated in cerebrospinal fluid spaces

b. Low grade astrocytomas must be differentiated from reactive gliosis

(1) **Reactive gliosis** — occurs in response to central nervous system injury; identifiable as increased cellularity of evenly-distributed, relatively uniform astrocytes; unless

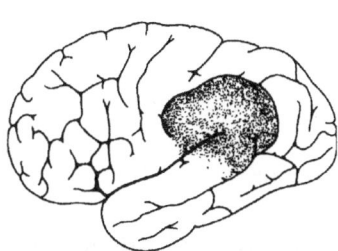

Gyral enlargement by astrocytoma.

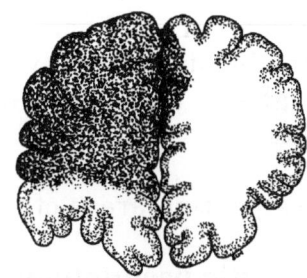

Diffusely-infiltrating cerebral astrocytoma.

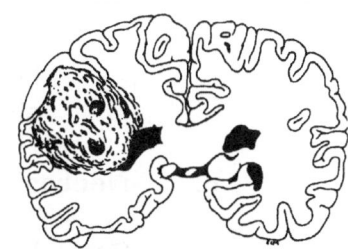

Cerebral glioblastoma multiforme.

injured, gray-white junction is preserved; ordinarily there are no mitotic figures

(2) **Low grade astrocytoma** — increased cellularity of irregularly-distributed, pleomorphic (varying size and shape) astrocytic cells with variable nuclei (nuclear atypia); gray matter-white matter junction is blurred or lost; mitotic figures identifiable (although infrequent)

    (a) Fibrillary astrocytes — cells with numerous long hyaline eosinophilic processes containing sheaves of glial intermediate filaments (glial fibrillary acidic protein, GFAP)

    (b) **Gemistocytic astrocytes** — cells with abundant perinuclear hyaline eosinophilic cytoplasm containing masses of glial intermediate filaments

c. **Endothelial hyperplasia**

(1) Astrocytomas produce vasoproliferative factor that results in marked hypertrophy and hyperplasia of endothelial cells with formation of thick irregular vascular profiles ("glomeruloid structures" that mimic renal glomeruli)

(2) **Gliosarcoma** — in high grade astrocytoma (glioblastoma multiforme), proliferating cells of blood vessel walls undergo malignant transformation to sarcoma resulting in mixed tumor with both glial and sarcoma elements

Perinecrotic pseudopalisading and glomeruloid structures of endothelial hyperplasia in glioblastoma multiforme.

d. **Necrosis** — serpentine areas of necrotic tumor

(1) **Perinecrotic pseudopalisading** — peripheral alignment of nuclei in rows around areas of necrosis

e. **Juvenile cystic pilocytic astrocytoma**

    (1)   Low grade astrocytic tumor usually occurring in children and involving cerebellum, optic nerve or tract, or hypothalamus

    (2)   Features include numerous elongated astrocytic processes, Rosenthal fibers, and accumulation of intercellular matrix material

        (a)   **Rosenthal fiber** — densely eosinophilic club-shaped condensation of protein in astrocytic process; presumably originates from glial intermediate filament protein (glial fibrillary acid protein)

        (b)   Accumulated intercellular matrix material results in microcystic appearance; coalescence of smaller cysts produces macrocysts

f. Specific varieties of astrocytoma:

    (1)   **Anaplastic astrocytoma** — malignant astrocytoma **without necrosis**; survival approaches 2 years

    (2)   **Glioblastoma multiforme** — malignant astrocytoma with histologic features of **necrosis**; survival 6 to 12 months

    (3)   **Optic glioma** — tumor involving optic nerve, chiasm, or tract (and sometimes infiltrating hypothalamus) often occurring (mostly in children) as part of **neurofibromatosis** (von Recklinghausen's syndrome)

    (4)   **Subependymal giant cell astrocytoma** — characteristic of **tuberous sclerosis**

        (a)   Subependymal cerebral tumors composed of unusual very large astrocytic cells with abundant eosinophilic cytoplasm and large irregular nuclei with prominent nucleoli (these cells have some histologic resemblance to neurons)

        (b)   Identifiable radiologically as **multiple subependymal nodules** (termed "candle gutterings" for resemblance to candle wax drippings)

    (c)   Other features of tuberous sclerosis include **mental retardation, seizures** (particularly infantile spasms), and hamartomas of various organs including cerebral cortical hamartoma ("**cortical tuber**"), facial angiofibroma (**adenoma sebaceum**), cardiac rhabdomyoma, renal angiomyolipoma, **shagreen skin patch**, retinal glial hamartoma, and hypomelanotic skin macules ("ash leaf spots")

    (5)   **Butterfly glioma** — growth of malignant astrocytoma through **corpus callosum** to involve opposite cerebral hemisphere resulting in "butterfly" appearance on radiologic studies

    (6)   **Gliomatosis cerebri** — bilateral diffuse involvement of cerebral white matter (nearly whole centrum semi-ovale)

3.   Prognosis is generally poor

    a.   In adults, treatment involves **surgical debulking** followed by **radiation therapy**

    b.   In children, brain stem gliomas (usually of pons) identified by radiologic imaging studies are often treated with radiation therapy without surgical debulking

    c.   Spinal cord tumors — extensive surgical debulking can prolong survival

    d.   Astrocytomas of **cerebellum or optic nerve in children** and of **spinal cord in adults** are **low-grade and slow growing**, resulting in extended survival

B.   **Oligodendroglioma**

1.   Slow-growing, **infiltrative** tumor of **malignant oligodendrocytes**, usually located in cerebral hemisphere (frequently in **temporal lobe**)

2.   Often associated with history of **chronic seizure disorder (complex partial seizure disorder or temporal lobe epilepsy)** that becomes progressively more refractory to anticonvulsant medications

3. **Calcification** is often visible on radiologic imaging studies

4. Histologic appearance of increased cellularity of relatively **uniform malignant oligodendrocytes**

    a. Cells have characteristic round to oval dark nucleus with **clear perinuclear halo ("fried egg appearance")**; this perinuclear clearing is artifactual from autolysis prior to formaldehyde fixation

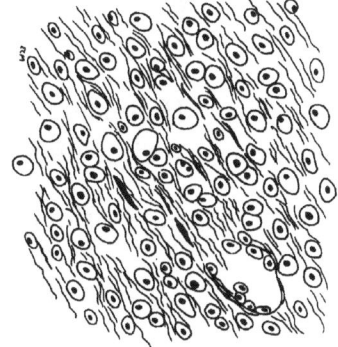

Oligodendroglioma with characteristic "fried-egg" appearance.

    b. **Thin-walled tumor vasculature ("chicken-wire vessels")** separates groups of tumor cells

    c. Usually **intermixed with malignant astrocytes**; prognosis is based on extent of astrocytic component, with lower percentage of astrocytes correlated with longer survival

5. Treatment involves extensive surgical resection followed by radiation therapy

C. **Ependymoma**

1. Slow-growing tumor of malignant **ependymal cells**

2. Clinical presentation

    a. Common in childhood (usually arising in fourth ventricle) or in young adults (arising in cauda equina)

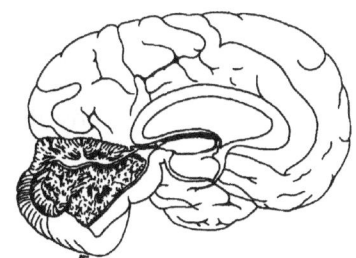

Ependymoma filling fourth ventricle.

    b. Typically poor survival except for those in cauda equina which can be completely resected

    c. **Intracranial tumors** — initial presentation with **hydrocephalus** and increased intracranial pressure due to ventricular obstruction

    d.   **Spinal tumors** produce **backache** and **lower motor neuron signs**

3.   Histologic appearance of sheets of relatively uniform cells with eosinophilic cytoplasm containing glial intermediate filaments

    a.   **Perivascular pseudorosette — perivascular** cellular arrangement with ependymal cell processes extending to vessel wall; mimics tanycytes (anterior third ventricular ependymal cells that have processes extending to hypothalamic blood vessels)

    b.   **Flexner-Wintersteiner (true ependymal) rosette** — circular arrangement of cells with processes forming small **lumen** (sometimes with cilia extending into lumen)

    c.   **Syrinx — slit-like cavity of spinal cord**; tumor often forms mural nodule

    d.   **Papillary ependymoma** — tumor cells arranged in **papillary fronds** (can be confused with choroid plexus papilloma)

    e.   **Subependymoma** — tumor arising from floor of fourth ventricle composed of small nests of uniform nuclei surrounded by large areas of interweaving hyaline eosinophilic processes containing glial intermediate filaments

    f.   **Myxopapillary ependymoma** — occurs at lower end of spinal cord and is characterized by extensive **mucin accumulation**; can involve presacral or retrosacral tissues

Perivascular pseudorosette.

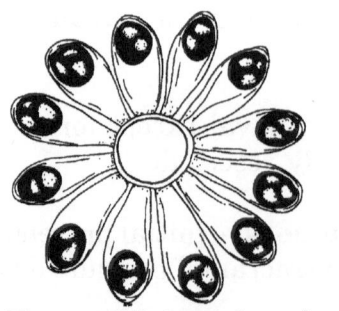

Flexner-Wintersteiner (true) rosette.

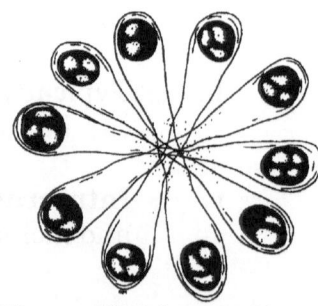

Homer-Wright rosette.

4. Treatment involves surgical resection; prognosis relates to degree of completeness of surgical resection, with long survival in some cases; radiation therapy can be used in cases of incomplete surgical resection

IV. **Primitive neuroectodermal tumor (PNET)**

A. General features

1. Highly malignant tumor composed mostly of **undifferentiated neuroectodermal cells** (arising from primitive germinal neuroepithelial cells)

   a. Mostly small cells with **dark (hyperchromatic) nuclei** and **minimal perinuclear cytoplasm**; relative lack of eosin-staining cytoplasm and prominence of hematoxylin-staining nuclei (in standard hematoxylin and eosin stained slides) results in "**blue cell tumor**" appearance

   b. **Rapid growth** with **numerous mitotic figures**

   c. Many individual degenerating cells and larger areas of **necrosis**

   d. Spread along CSF pathways results in **metastases throughout neuraxis**

   e. Treatment involves aggressive surgical debulking, followed by radiation therapy (tumor is very radiosensitive) to whole neuraxis (brain and spine) plus chemotherapy; aggressive therapy has produced many long-term survivors; treatment must be modified in children less than 2 years of age due to excessive sensitivity of normal tissue to radiation damage

2. Occasional evidence of differentiation along glial, neuronal, or other lines

   a. **Perivascular pseudorosette** — suggestive of ependymal (tanycyte) differentiation

   b. **Flexner-Wintersteiner rosette** — suggestive of neural tube, ventricular (ependymal), or optic vesicle (retinal) differentiation

   c. **Homer-Wright rosette** — circular arrangement of cells around fibrillary core (mimicking neuronal processes) suggestive of neuronal differentiation

d.   **Fleurette** — abortive photoreceptor structures suggestive of retinal differentiation

e.   **Neurosecretory granules** — vesicles containing neurotransmitter suggestive of neuronal differentiation

f.   Rare striated or smooth muscle (myoblastic) or melanocytic differentiation

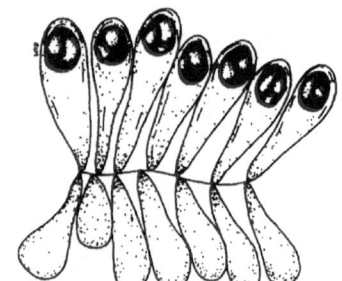

Fleurette.

B.   Tumor designated by site and by extent of differentiation

1.   **Medulloblastoma**

   a.   Common childhood tumor of **cerebellum**

      (1)   Arises in **midline cerebellum** (vermis) and fills fourth ventricle with resultant hydrocephalus (from ventricular obstruction) and increased intracranial pressure

      (2)   In adolescents or young adults, tumor tends to arise in cerebellar hemisphere and incites intense arachnoidal fibrosis (**desmoplastic medulloblastoma**)

   b.   Clinical symptoms — **increased intracranial pressure**, with headache, vomiting, papilledema, and lethargy

2.   **Cerebral neuroblastoma** — tumor arising in **cerebral hemispheres**

3.   **Retinoblastoma**

   a.   Common childhood tumor arising in **retina**; often bilateral and **familial**

   b.   Treatment involves **enucleation;** cure is possible if tumor has not spread into optic nerve; however, familial cases have predisposition to **second malignancies** (particularly **osteosarcoma** of femur)

4.   **Pineoblastoma** or **pineocytoma** — **pineal gland** tumor; pineocytoma tends to have more differentiated cells containing neurosecretory granules

V.   Peripheral **neuroblastoma**

    A.   Common, **highly-malignant tumor** of childhood originating from **adrenal medulla** or from sympathetic ganglia (usually retroperitoneal, but occasionally mediastinal)

    B.   Clinical symptoms

        1.   Generally presents as **abdominal mass** detected on routine pediatric evaluation

        2.   **Opsoclonus-myoclonus** ("dancing eyes and dancing feet") — paraneoplastic syndrome (nonmetastatic distant effect of tumor) consisting of rapid, involuntary, repetitive conjugate eye movements (in all directions), associated with ataxia and myoclonus

    C.   Histologic appearance similar to intracranial primitive neuroectodermal tumor

    D.   Treatment involves resection of primary tumor mass followed by chemotherapy or radiation therapy; blood-born metastases are common and prognosis is poor

VI.   **Pheochromocytoma**

    A.   Benign tumor in adults originating from adrenal medullary chromaffin cells

    B.   Clinical symptoms

        1.   Intermittent attacks of **diaphoresis** (profuse sweating), **tachycardia**, and **headaches** associated with **markedly elevated blood pressure**

        2.   Diagnosis established by demonstrating **excessive levels of urinary catecholamines** due to **tumor secretion of epinephrine and norepinephrine**

    C.   Histologic appearance of large cells filled with numerous neurosecretory granules containing epinephrine or norepinephrine

    D.   Treatment involves complete surgical excision

VII.   **Craniopharyngioma**

    A.   Slowly-growing **cystic tumor** in suprasellar region presenting usually in childhood or adolescence

B.   Clinical symptoms

    1.   **Visual field abnormalities** — compression of optic chiasm resulting in bitemporal hemianopsia; pressure on optic nerve or tract can produce optic atrophy or other visual field defects

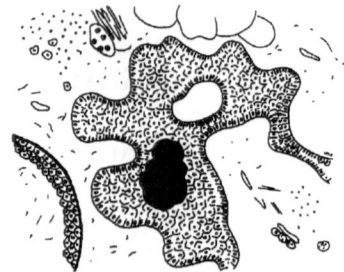

    2.   **Endocrine disturbances** — **hyperprolactinemia** from compression of pituitary stalk or hypothalamus; other endocrine abnormalities can also occur

    3.   Extension into third ventricle can cause obstructive hydrocephalus and increased intracranial pressure

Craniopharyngioma with squamous cell proliferation, connective tissue, and squamous debris.

C.   Calcification is often identifiable on radiologic studies

D.   Arises from **squamous cell rests** derived from **Rathke's pouch**

    1.   Histologic appearance of nonkeratinizing squamous epithelium often resembling primitive tooth enamel organ (adamantinomatous pattern)

    2.   Cyst is filled with thick "machinery oil-like" fluid containing abundant cholesterol crystals; leakage of this fluid into subarachnoid space elicits severe chemical meningitis

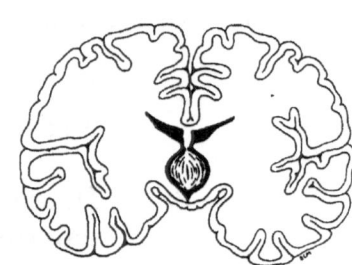

E.   Treatment involves complete surgical excision

VIII.   **Colloid cyst of third ventricle**

A.   Benign cystic tumor arising in **anterior roof of third ventricle**

Colloid cyst of third ventricle.

B.   Clinical symptoms of intermittent severe bifrontal **headache that changes with head position** (worse when lying on back, improved by sitting upright or bending forward)

    1.   Due to **intermittent obstruction** ("ball valve" effect) of cerebrospinal fluid flow through foramina of Monro

2. Can present with acute coma (from increased intracranial pressure) leading rapidly to death

C. Histologic appearance of mucin-containing cyst lined by pseudostratified ciliated columnar epithelium

D. For tumors demonstrated during life (by radiologic imaging studies), urgent ventricular decompression and neurosurgical excision is curative

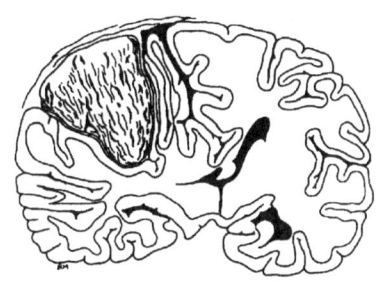

Meningioma attached to dura and indenting cerebral hemisphere.

IX. Meningioma

A. Relatively common, **slow-growing, benign tumor** arising from arachnoidal (leptomeningeal) cap cells (cells of arachnoid villi) usually presenting in older patients; small tumors often found incidentally at autopsy

1. Attached to dura with intracranial vascular supply from external carotid artery (cerebral convexity tumor often supplied by middle meningeal artery)

Whorls in meningioma.

2. Tendency to infiltrate dura and bone with marked thickening of bone (hyperostosis)

3. En plaque meningioma — tumor growth as flattened sheet, particularly along sphenoid ridge

4. Multiple meningiomas are associated with neurofibromatosis

5. Intraventricular meningiomas arise along choroidal fissure in association with choroid plexus

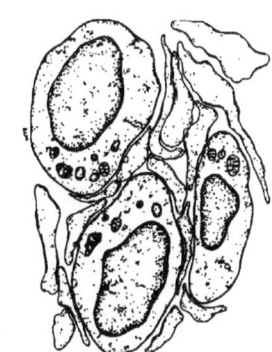

6. Tumor indents and compresses (but does not invade) surrounding neural tissue

Cells in meningioma have interdigitating processes and no basal lamina.

B. **More frequent in women** than men; tumor cells can express estrogen, progesterone, or androgen receptors

C. Clinical symptoms relate to central nervous system compression; because of slow growth, tumor can become very large before significant clinical symptoms develop

D. Histologic appearance of uniform cells with indistinct cytoplasmic boundaries (syncytial appearance) due to numerous interdigitating cell processes

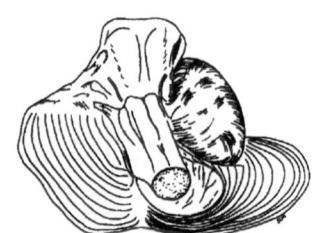

Acoustic neuroma growing in cerebellopontine angle producing cerebellar and brain stem compression.

   1. Whorls — cells arranged in concentric lamellae, which become hyalinized and then calcified (psammoma bodies)

   2. Tumors with poor prognosis (high likelihood of recurrence) have identifiable mitotic figures and finger-like projections into underlying brain

E. Treatment is complete surgical excision

X. **Schwannoma (neurilemmoma)**

A. Slow-growing, encapsulated benign **tumor of Schwann cells involving peripheral or cranial nerves** in any location

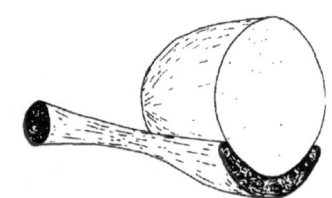

Schwannoma growth compresses adjacent nerve fascicles.

   1. Most common **primary intraspinal tumor**

   2. **"Acoustic neuroma"** — most frequent intracranial site for schwannoma is on **cranial nerve VIII**

B. Tumor growth compresses and displaces nerve fascicles peripherally resulting in progressive nerve dysfunction; tumor can be peeled away from nerve during surgical removal

C. Biphasic histologic pattern of intermixed **compact areas (Antoni A)** and **loose microcystic areas (Antoni B)**

   1. **Verocay body** — alignment (palisading) of elongated nuclei on either side of eosinophilic fibrillary area

2. Basal lamina (basement membrane) completely surrounds each cell

3. Long-spacing collagen — unique collagen banding pattern characteristic of schwannoma

D. Treatment involves complete surgical excision

E. **Neurofibromatosis** — autosomal dominant **neurocutaneous syndrome (phakomatosis)**

Nuclear palisading (Verocay body) in schwannoma.

1. **Neurofibromatosis type 1** (NF-1; peripheral or classical neurofibromatosis; **von Recklinghausen's syndrome)** — multiple **café-au-lait spots, Lisch nodules** (iris hamartomas appearing as small yellow or brown elevations), cutaneous neurofibromas, **spinal or cranial nerve root neurofibromas or schwannomas,** and skeletal anomalies; gene localized to chromosome 17

2. **Neurofibromatosis type 2** (NF-2; **central neurofibromatosis)** — **bilateral acoustic neuromas (schwannomas)** with few café-au-lait spots; gene localized to chromosome 22

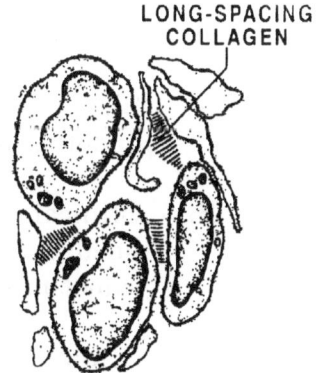

Long-spacing collagen identified by electron microscopy in schwannoma.

XI. **Neurofibroma**

A. Slow-growing benign **tumor of Schwann cells** involving peripheral or cranial nerves, usually affecting **terminal cutaneous nerve branches** with only clinical symptom being **palpable subcutaneous mass**

1. Tumor growth produces **fusiform expansion of nerve**, incorporating axons and other nerve elements into tumor mass; tumor infiltrates nerve distant from main tumor mass, necessitating wide surgical excision (transection of nerve at some distance proximally and distally in order to remove all tumor)

2. Tumor involvement of large nerve trunk results in rope-like ("bag of worms") appearance

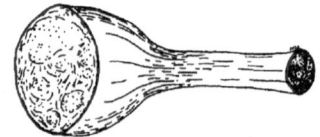

Neurofibroma produces fusiform expansion of nerve, incorporating nerve elements into tumor mass.

   a. Such tumors usually occur as manifestation of neurofibromatosis and have propensity for malignant transformation

   b. Dumbbell or hourglass tumor — growth of **spinal nerve root neurofibroma** through intervertebral (neural) foramen

B. Histologic appearance of mucinous matrix containing loose interlacing bundles of elongated spindle cells with wavy nuclei; axonal processes can often be identified within tumor mass

C. Treatment involves complete surgical excision

XII. **Hemangioblastoma** of cerebellum (**Lindau's tumor**)

A. Slow-growing, cystic, highly vascular, benign tumor arising from capillary endothelium

B. Clinical presentation of cerebellar ataxia, vertigo or dizziness, along with headache and papilledema or other signs of increased intracranial pressure; sometimes fatal due to cerebellar tonsillar herniation

C. Often associated with **polycythemia** due to tumor elaboration of erythropoietin-like factor

D. Histologic appearance of sheets of uniform **foamy cells (stomal cells)** interdigitated with **abundant capillary network**

E. **Von Hippel-Lindau disease** — about 10% of cerebellar hemangioblastomas are familial as part of syndrome characterized by:

1. Multiple intracranial (including cerebellar) or intraspinal hemangioblastomas

2. Retinal hemangioblastoma (**retinal angiomatosis** or **von Hippel tumor**)

3.   Renal and pancreatic cysts

4.   **Renal cell carcinoma**

F.   Treatment involves complete surgical excision

XIII.   Chordoma

A.   Slow-growing, indolent tumor of axial skeleton arising from remnants of embryonic **notochord**, usually in **sacrum** or in **clivus** at skull base

B.   Clinical symptoms arise from compression of adjacent neural tissue

1.   Clivus chordoma produces signs of cranial nerve, hypothalamic, and pituitary dysfunction

2.   Sacral chordoma produces back pain, urinary and anal sphincter dysfunction, and spinal root symptoms

C.   Histologic appearance of large mucin-containing, vacuolated (bubbly) **physaliphorous cells** in mucinous matrix

D.   Treatment involves complete surgical excision; however, since tumor is often inaccessible, removal is frequently incomplete resulting in recurrence

XIV.   Primary central nervous system lymphoma

A.   Occurs in individuals with **immune compromise** (as in **AIDS**) or in older individuals (generally over age 60 years) with no evidence of immunologic disorder

B.   **Deep cerebral hemisphere** tumor, often **bilateral** and showing homogenous contrast enhancement on radiologic imaging studies

C.   Histologic appearance of sheets of **malignant lymphoid cells accumulated around blood vessels**; associated with characteristic concentric reticulin pattern; usually classified as **high grade** lymphoma, often **immunoblastic (B-cell)** type

D.   Initial dramatic response to **corticosteroids** and **radiation** ("disappearing tumor") is generally followed by recurrence (including systemic involvement) and fatality within 3 years

XV. Metastatic tumors

A. Parenchymal brain metastases

1. **Most common central nervous system tumor**, affecting almost 20% of patients dying of cancer; may be single or multiple

2. Any malignancy can potentially metastasize to central nervous system, but most common are **lung cancer, breast cancer,** or **malignant melanoma**

3. Frequency of metastatic site is roughly proportional to size of brain region (cerebrum > cerebellum > brain stem)

4. Metastatic tumors tend to push into brain (instead of diffusely infiltrating like gliomas) producing what is often mistaken for well-defined margin; surrounding brain is usually **massively edematous**

5. Clinical presentation includes **headache, papilledema,** and signs of **increased intracranial pressure**; focal signs depend upon specific site of nervous system involvement

6. Treatment depends on location and number of metastases, type of primary tumor, degree of systemic involvement, and presumed general prognosis

   a. **High-dose corticosteroids** reduce edema and can dramatically improve symptoms (particularly headache and signs of increased intracranial pressure)

   b. **Surgical resection** indicated for solitary, readily accessible brain lesion or lesion involving spinal cord

   c. **Radiation therapy** indicated for multiple metastases or when patient is not surgical candidate because of systemic condition

B. Meningeal carcinomatosis, lymphoma, or leukemia

1. Common **complication of cancer**, particularly lymphoma, leukemia, small-cell anaplastic ("oat-cell") lung carcinoma, breast carcinoma, "signet-ring cell" gastric carcinoma, or choriocarcinoma

2. Clinical presentation of **meningeal irritation** (backache, headache, and stiff neck) along with peripheral or cranial nerve signs (cranial nerve palsies or radicular pain and weakness) due to focal cranial nerve or spinal root infiltration

3. **Carcinomatous meningitis — cerebrospinal fluid examination** reveals malignant cells, along with inflammatory cells (lymphocytes and neutrophils), elevated protein level, and decreased glucose level

4. Treatment involves radiation therapy to whole neuraxis (craniospinal radiation) and injection of chemotherapeutic drugs directly into subarachnoid space (intrathecal chemotherapy); prognosis is poor since meningeal involvement usually indicates widespread metastatic disease

C. **Spinal metastases**

1. **Local tumor spread to epidural space** by direct extension from involved vertebral body or through intervertebral foramina; most common in thoracic and lumbar vertebrae

2. Initial symptoms include **unremitting** severe localized back or radicular **pain**, weakness or numbness in legs, and urinary retention; rapidly progresses to complete **paraplegia** (from compression of anterior spinal artery with consequent spinal cord infarction)

3. Treatment involves administration of **high-dose corticosteroids** to reduce edema and surgical decompression or radiotherapy to reduce tumor bulk and relieve pressure on spinal cord

SUGGESTED ADDITIONAL READING

Burger PC, Scheithauer BW: *Tumors of the Central Nervous System. Atlas of Tumor Pathology, Third Series, Fascicle 10.* Washington, D.C., Armed Forces Institute of Pathology, 1994.

Burger PC, Scheithauer BW, Vogel FS: *Surgical Pathology of the Nervous System and Its Coverings,* 3rd ed. New York, Churchill Livingstone Inc, 1991.

Harkin JC, Reed RJ: *Tumors of the Peripheral Nervous System. Atlas of Tumor Pathology. Second Series, Fascicle 3.* Washington, D.C., Armed Forces Institute of Pathology, 1969.

Kleihues P, Burger PC, Scheithauer BW: *Histological Typing of Tumours of the Central Nervous System,* 2nd ed. Berlin, Springer-Verlag, 1993.

Reed RJ, Harkin JC: *Supplement. Tumors of the Peripheral Nervous System. Atlas of Tumor Pathology. Second Series, Fascicle 3*. Washington, D.C., Armed Forces Institute of Pathology, 1982.

Rubinstein LJ: *Supplement. Tumors of the Central Nervous System. Atlas of Tumor Pathology. Second Series, Fascicle 6*. Washington, D.C., Armed Forces Institute of Pathology, 1982.

Rubinstein LJ: *Tumors of the Central Nervous System. Atlas of Tumor Pathology. Second Series, Fascicle 6*. Washington, D.C., Armed Forces Institute of Pathology, 1972.

Russell DS, Rubinstein LJ: *Pathology of Tumors of the Nervous System*, 5th ed. Baltimore, Williams & Wilkins, 1989.

CHAPTER 8: PATHOLOGY OF CRANIOCEREBRAL TRAUMA

I. **Craniocerebral trauma** — soft, easily disrupted brain is protected by rigid skull, underlying dense fibrous dura, and cushioning layer of cerebrospinal fluid; nonetheless, excessive agitation of cranium or damage to any of these layers can produce brain injury:

   A. **Missile injury** — damage produced by moving object striking cranium; most often refers to bullet injury

   B. **Penetrating (open) head injury** — disruption (penetration) of cranial vault with opening through skin and cranial bones to expose damaged brain; most often associated with missile injury, but also occurs with stab wounds, or blows by larger moving objects (such as flying fragments or falling objects)

   C. **Closed (non-penetrating) head injury** — damage to brain without disruption of skin over cranial vault; most often results from blunt trauma

   D. **Acceleration/deceleration injury** — damage produced by movement of brain within confines of cranial vault; brain injury results from tearing during violent movement or from impact of striking interior of skull or dural folds

   E. **Compressive injury** — damage resulting from compression of cranial vault with resultant fracture of bones and injury to underlying brain

II. **Missile injuries**

   A. Damage relates to various factors

      1. Composition — soft bullets fragment resulting in multiple paths of destruction through brain

      2. Size and shape — larger bullets do greater damage

3.   Velocity — higher velocity bullets have more energy (energy is proportional to mass and to square of velocity) to dissipate in passing through skull and brain resulting in greater explosive effect on tissues

4.   Line of flight and direction of impact — different paths can produce strikingly different degrees of brain injury

   a.   Superficial (tangential) path — injury varies from only superficial scalp and soft tissue injury to depressed skull fractures with laceration of underlying brain

   b.   **Penetrating injuries** — skin, hair, bone, and missile fragments are driven into brain

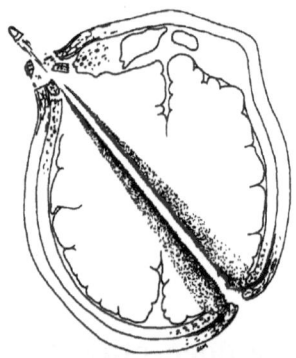

Through-and-through bullet injury driving bone fragments into brain at entry site and exploding outward at exit site.

      (1)   Low velocity missiles (less than 1100 ft/sec) **ricochet on inner table of skull**, thereby widening area of damage

      (2)   **Through-and-through injuries** — usually associated with high velocity missiles; skin, hair, and bone fragments driven into brain; **shock waves** stretch, shear, or rupture blood vessels, nerves, and bone at considerable distance from missile entry or exit site

   c.   Bullet path through frontal poles often produces only cognitive deficits, while path through brain stem is immediately fatal

B.   Damaging consequences

1.   Laceration — track of bullet creates channel of necrotic tissue damaged by direct effect of bullet

2.   Explosive gases — expansion of hot gases trailing behind high velocity bullet has explosive effect in brain causing widespread damage

3.   Hemorrhage — extent of bleeding depends on site of blood vessel injury (damage to dural sinus or vessels of circle of Willis can result in massive hemorrhage) and time over which bleeding occurs (little hemorrhage occurs with immediate death)

4. Cerebral edema — edema resulting from local injury is combined with edema from hypoxia or ischemia due to shock or breathing difficulties

III. **Closed head injury**

A. **Scalp injuries** — scalp lesions may be only reliable evidence of underlying injuries to skull or brain

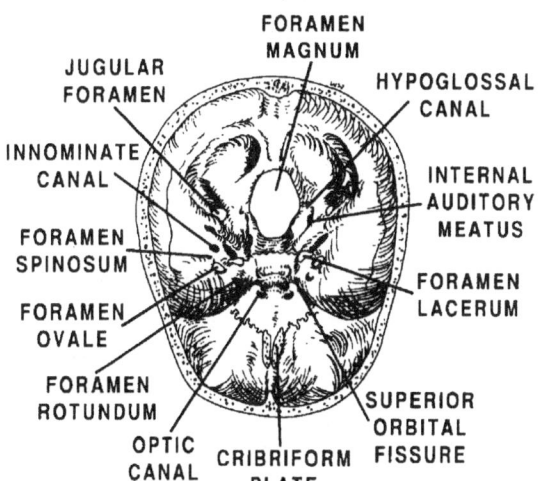

Skull base foramina containing structures often damaged in basilar skull fracture.

B. **Skull fracture**

1. Fracture types

    a. **Depressed skull fracture** — bone **displaced** inward toward brain

    b. Linear (fissure) skull fracture — breaking (bursting) of skull bones due to bending at site somewhat distant from point of impact

    c. Diastatic skull fracture — linear skull fracture separating cranial sutures; occurs in children before completion of osseus obliteration of sutures

    d. **Compound skull fracture — laceration of scalp** associated with skull fracture

    e. **Growing skull fracture** — expanding linear fracture (usually in young children) associated with dural tear allowing leptomeninges to herniate into fracture site and expand with cerebrospinal fluid pressure

2. **Basilar skull fracture** — fracture through base of skull

    a. Often passes through skull foramina, damaging cranial nerves or blood vessels that normally traverse those foramina

        (1) Cribriform plate — olfactory nerves

        (2) Optic canal — optic nerve

(3)   Superior orbital fissure — oculomotor nerve, trochlear nerve, abducens nerve, opthalmic division of trigeminal nerve

(4)   Foramen rotundum — maxillary division of trigeminal nerve

(5)   Foramen ovale — mandibular division of trigeminal nerve

(6)   Foramen lacerum — internal carotid artery and sympathetic nerve fibers

(7)   Foramen spinosum — middle meningeal artery and vein

(8)   Innominate canal — lesser superficial petrosal nerve

(9)   Internal auditory meatus — facial nerve, vestibulocochlear nerve

(10) Jugular foramen — glossopharyngeal nerve, vagus nerve, spinal accessory nerve, internal jugular vein

(11) Hypoglossal canal — hypoglossal nerve

(12) Foramen magnum — cervicomedullary junction, spinal accessory nerve, vertebral arteries, anterior spinal arteries

b.   **Battle's sign**

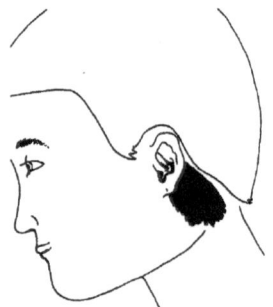

(1)   Fractures through **petrous bone (ear)** producing **subcutaneous blood over mastoid**

(2)   Usually associated with cranial nerve VIII damage and tympanic membrane rupture with blood drainage or CSF leakage into middle ear or out external auditory canal (**CSF otorrhea**)

Battle's sign: blood in soft tissue overlying mastoid.

c.   **"Raccoon eyes"**

(1)   Fracture through **anterior cranial fossa** resulting in **subcutaneous blood around eyes**

(2) Often associated with CSF leakage into sinuses (**CSF rhinorrhea**) and damage to olfactory nerve (cranial nerve I), optic nerve (cranial nerve II), or oculomotor nerve (cranial nerve III)

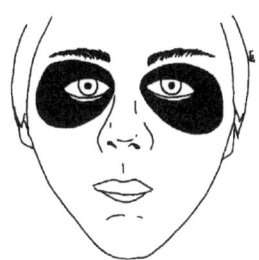

Raccoon eyes: blood in periorbital soft tissues.

d. Carotid-cavernous fistula — fractures through cavernous sinus associated with **tearing** of **carotid artery within cavernous sinus**, resulting in massive **shunting** of blood producing venous distention; presents as **painful, pulsating exophthalmos**; intracavernous cranial nerves can also be damaged

e. Cerebrospinal fluid leakage and pneumocele — fractures through inner wall of nasal sinuses or mastoid air cells permit entry into cranial cavity of air (visible on radiologic imaging studies) and bacteria (which can produce meningitis or brain abscess); persistent CSF leakage requires surgical repair to prevent recurrent infection

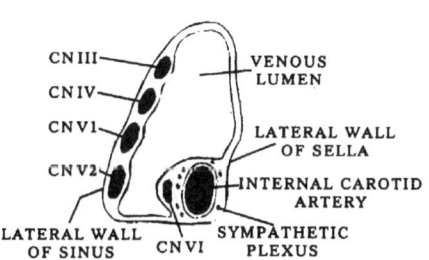

Cavernous sinus structures.

3. Cranial vault fractures — can be asymptomatic or result in middle meningeal artery injury (with consequent epidural hematoma) or venous sinus tear (with consequent acute subdural hematoma)

a. **Epidural hematoma — arterial** hemorrhage from **temporal bone fracture tearing middle meningeal artery** which results in brisk arterial (high pressure) bleeding, creating rapidly expanding mass **between dura and cranial bone** (hence, *epidural*)

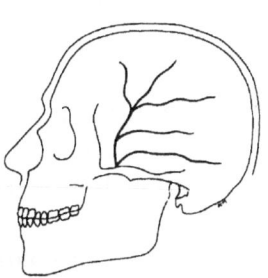

Course of middle meningeal artery along interior lateral skull.

(1) Clinical symptoms — initial unconsciousness (due to cerebral concussion); usually followed by arousal (**"lucid interval"**), although up to 50% of patients may not arouse due to severe head injury; then followed by progressive **headache** and **drowsiness, hemiparesis, dilating pupil** on side of hemorrhage (indicating **transtentorial uncal herniation**

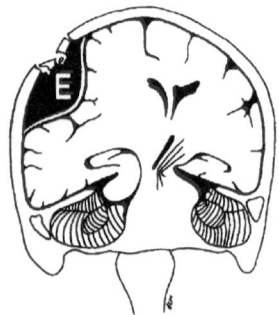

Epidural hematoma.

with compression of oculomotor nerve), and slowing of pulse and respirations with increase in blood pressure (**Cushing's reflex** indicating brain stem distortion with medullary dysfunction from herniation)

(2) Death ensues unless bleeding is promptly controlled and blood clot removed; requires immediate neurosurgical evacuation of clot, which often produces complete recovery; survivors of delayed intervention frequently have permanent brain damage

b. **Acute subdural hematoma** — damage to **venous sinus** results in rapid accumulation of blood between arachnoid and dura (hence, subdural) with clinical presentation mimicking epidural hematoma and necessitating prompt neurosurgical evacuation and control of bleeding

C. **Chronic subdural hematoma** — **venous** hemorrhage from damage to veins communicating between cerebral cortex and venous sinuses (**"bridging veins"**) resulting in accumulation of venous (low pressure) blood **between arachnoid and dura** (hence, *subdural*)

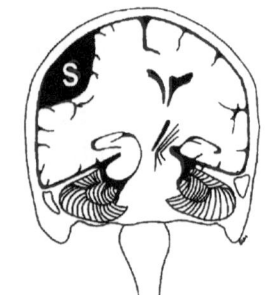

Subdural hematoma.

1. More common than epidural hematoma, since veins can be torn with only minimal or incidental trauma (without fracture)

2. Most often located over lateral surface (convexity) of frontal and parietal lobes

3. **Blood accumulates slowly**; bleeding ceases spontaneously (probably after few hours) due to tamponade effect

4. Slowly progressive clinical symptoms of headache, confusion, hemiparesis, apathy, lethargy, and ultimately coma

5. Radiologic imaging studies confirm diagnosis; MRI preferable, since although initial CT scans can show hyper-density of hematoma (denser appearance than adjacent brain), subsequent isodensity (similar appearance to adjacent brain) makes detection difficult

6. Composition in relation to time of injury:

    a. Acute (within 48 hours) — dark red semi-liquid blood

    b. Subacute (from 2 days to 2 weeks) — red-black gelatinous mixture of liquid and clotted blood

    c. Chronic (greater than 2 weeks) — dark turbid orange-brown fluid

7. Response to subdural hemorrhage is organization with formation of delicate fibrous **neomembrane** that is first apparent on dural side at about 1 week and completely encloses clot by 2 weeks

8. Hematomas can undergo **gradual enlargement** over several weeks due to abnormal permeability of vessels in neomembrane, rupture of delicate vessels in neomembrane with **recurrent bleeding** ("rebleeding"), and increased osmotic pressure from blood breakdown products; such enlargement can produce symptoms of brain compression

9. Pathologic features

    a. **1 week — thin neomembrane** composed of about 12 layers of fibroblasts visible on dural surface

    b. **2 weeks** — complete **neomembrane surrounding partially liquified clot**

    c. **3 weeks — liquified clot**; vessels permeating fluid space; numerous hemosiderin-containing macrophages; thick neomembrane

      d.   **1-3 months** — vascular neomembrane with beginning hyalinization; **large delicate vessels with frequent rebleeding** (fresh hemorrhage)

      e.   **3-6 months — hyalinized neomembrane**

      f.   **1 year — neomembrane resembling dura**

10.   Infants or children can develop subdural hematomas from falls or from child abuse

      a.   Clinical presentation — increased intracranial pressure with enlarging head (increased head circumference), bulging fontanelle, vomiting, irritability, and lethargy; often associated with seizures

      b.   In child abuse, fractures of skull and other bones are often evident; retinal hemorrhages without evidence of external trauma suggests **"shaken baby syndrome"**

11.   Small subdural hematomas can be clinically insignificant and slowly absorbed, but larger hematomas require neurosurgical drainage

D. **Brain injury**

1.   **Concussion**

      a.   **Temporary** impairment of cerebral neuronal function following head injury, without evidence of structural damage

      b.   Can occur as only consequence of head injury or can accompany more severe injuries (such as skull fractures and epidural hematoma)

2.   **Contusion (bruising) and laceration (tearing) of brain tissue**

      a.   Results from physical distortion of brain tissue and is associated with necrosis, hemorrhage, and edema

      b.   Distribution of surface lesions

          (1)   Injury to tips of gyri (sparing depths of sulci) with resultant wedge-shaped hemorrhages (apex directed into

white matter) often visible on radiologic imaging studies (particularly MRI)

(2) **"Coup" lesion** — focal damage under site of impact

(3) **"Contrecoup" lesion** — focal damage at distance (usually opposite side) from impact due to angular and rotatory movements of brain relative to skull

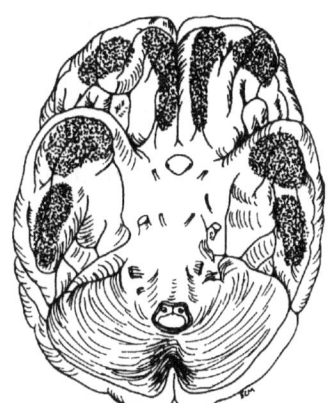

Location of contrecoup contusions on undersurfaces of frontal and anterior temporal lobes.

(a) **Undersurface of frontal lobe or anterior part of temporal lobe** — from brain impact against bony irregularities of orbital roof or lesser wing of sphenoid

(b) Mesial temporal lobe — from brain impact against sharp edges of tentorial incisura

(c) Corpus callosum — from brain impact against sharp inferior edge of falx cerebri

c. Large intracerebral hemorrhages or hemorrhages into deep nuclei (basal ganglia or thalamus) can also occur

d. **Diffuse axonal injury**

(1) Disruption of **axons in long tracts** attributable to shear and tensile strains from acceleration/deceleration

(2) Predilection for parasagittal cerebral white matter, corpus callosum, internal capsule, deep cerebellar white matter, and dorsolateral midbrain and pons (involvement of brain stem reticular formation results in coma)

(3) Initial histologic appearance of axonal retraction balls (axonal spheroids), with later evidence of wallerian degeneration and white matter gliosis

e.   **Plaque jaune** — brownish-yellow, glial-collagenous **scar** formed at sites of superficial **cortical contusions**

IV.   **Delayed complications** of head injury

A.   **Post-traumatic epilepsy** — result of contusion-laceration of cerebral cortex

    1.   Seizures following head injury may be classified as:

        a.   **Immediate (at time of injury** or within minutes of injury) — usually not associated with later development of epilepsy

        b.   **Early (within first week** following injury) — particularly common in children and usually associated with later epilepsy

        c.   **Late (after first week** following injury) — post-traumatic epilepsy; onset usually within 2 years following injury

    2.   Occurrence in about 5% of all patients with closed head injuries and up to 50% with depressed skull fracture or penetrating injuries

B.   **Postconcussion syndrome — headache, dizziness, and personality changes** following concussive head injury; most common in patients with pre-existing (prior to head trauma) psychiatric problems (such as depression)

C.   **Post-traumatic hydrocephalus** — arachnoidal fibrosis following resolution of subarachnoid hemorrhage obstructs cerebrospinal fluid pathways resulting in communicating hydrocephalus

D.   **Post-traumatic dementia (dementia pugilistica, "punch drunk syndrome")**

    1.   Repeated cerebral injuries (as occurs in boxers) results in syndrome of dementia, often developing many years after last injury

    2.   Clinical features are dementia (**memory disturbance, confusion, dysarthria**) and parkinsonian features (slow movements, shuffling wide-based gait, and tremor)

    3.   Pathologic findings include **thinning of corpus callosum**, lateral ventricular enlargement, **cavum septi pellucidi**, gliosis of inferior cerebellum, **loss of pigmented neurons** of substantia nigra and locus ceruleus, and presence of **neurofibrillary tangles** in cerebral cortical and brain stem neurons (**without any neuritic plaques**)

SUGGESTED ADDITIONAL READING

Drayer B (ed): Current concepts in imaging of craniofacial trauma. *Neuroimaging Clin NA* 1991; 1(2):259-524.

Jennett B, Teasdale G: *Management of Head Injuries*. Philadelphia, F. A. Davis Co, 1981.

Vinken PJ, Bruyn GW, Braakman R (eds): *Injuries of the Brain and Skull-Part I. Handbook of Clinical Neurology. Volume 23*. Amsterdam, North-Holland Publishing Co, 1975.

Vinken PJ, Bruyn GW, Braakman R (eds): *Injuries of the Brain and Skull-Part II. Handbook of Clinical Neurology. Volume 24*. Amsterdam, North-Holland Publishing Co, 1976.

Vinken PJ, Bruyn GW, Klawans HL, Braakman R (eds): *Head Injury. Handbook of Clinical Neurology. Volume 57*. Amsterdam, Elsevier Science Publishers, 1990.

CHAPTER 9: PERIPHERAL NERVE AND MUSCLE PATHOLOGY

I. Anatomy and physiology

    A. Peripheral nervous system, central nervous system, and neuromuscular system

        1. Peripheral nervous system includes spinal nerve roots and cranial nerve III though cranial nerve XII, dorsal root and cranial nerve ganglia, peripheral motor, sensory, and autonomic nerves, and specialized sensory end organs

        2. Central nervous system and peripheral nervous system are continuous

            a. Cell bodies of anterior horn neurons located in spinal cord (central nervous system) have axons located in ventral roots and peripheral nerves (peripheral nervous system)

            b. Dorsal root ganglion neurons send axons into spinal cord where they become part of central nervous system

        3. Myelin

            a. Peripheral nervous system — **Schwann cell produces only one internodal myelin sheath segment for one axon**

            b. Central nervous system — **oligodendrocyte produces many internodal myelin sheath segments on many different axons**

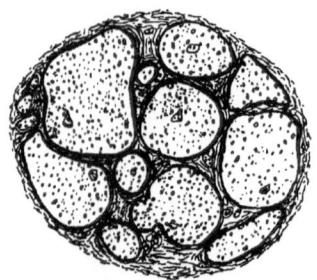

Peripheral nerve containing multiple nerve fascicles.

        4. Bundles of myelinated and unmyelinated nerve fibers form **fascicles** enveloped by

123

specialized connective tissue sheath (**perineurium**) continuous with pia-arachnoid which acts as **blood-nerve barrier** (analogous to central nervous system blood-brain barrier)

5.   Neuromuscular system consists of spinal cord anterior horn motor neuron, motor axon coursing in spinal root and peripheral nerve, neuromuscular junction, and skeletal muscle fiber

B.   Peripheral nervous system

1.   Axon extends from cell body in spinal cord (motor neuron in anterior horn) or in sensory ganglion

   a.   Axonal plasma membrane (axolemma) surrounds axonal cytoplasm (axoplasm) which contains mitochondria, neurofilaments, and neurotubules; mitochondria in myelinated fibers are particularly concentrated at nodes of Ranvier

   b.   Axonal function is dependent on metabolites transported from neuronal cell body (where protein synthesis occurs) by axoplasmic transport

NERVE FIBER DESIGNATION			DIAMETER	CONDUCTION VELOCITY	CONNECTIONS
ACTION	MYELINATION	TYPE	(μm)	(m/sec)	
Afferent	Myelinated	A-I	12-20	80-120	Muscle spindle primary endings; Golgi tendon organs
		A-II (Aß)	6-12	36-72	Muscle spindle secondary endings; Pacinian corpuscles; tactile receptors
		A-III (Aδ)	1-6	6-36	Pressure, tactile, temperature receptors
	Unmyelinated	C	1	0.5-2	Tactile, temperature, pain receptors
Efferent	Myelinated	Aα	12-20	80-120	Extrafusal striated muscle fibers
		Aγ	2-8	12-48	Intrafusal muscle fibers
		B	1-3	3-15	Preganglionic autonomic
	Unmyelinated	C	1	0.5-2	Postganglionic autonomic

c.   Axoplasmic transport

    (1)   Material movement between cell body and axon terminals, mediated by neurotubules (axonal microtubules) and neurofilaments

        (a)   Anterograde slow transport — movement of **cell organelles** such as lysosomes, mitochondria, and vesicles toward axon terminals at **rate of about 1 to 3 mm/day**

        (b)   Anterograde fast transport — movement of **peptides**, **proteins**, and **metabolites** toward axon terminals at rate of about 400 mm/day

        (c)   Retrograde transport — movement of materials from axon terminals toward cell body at varying rate of up to about 400 mm/day

    (2)   Pathology related to axonal transport

        (a)   Axonal ballooning (axonal spheroid) — axonal swelling due to constriction of axon with resultant accumulation of organelles and proteins proximal to constriction

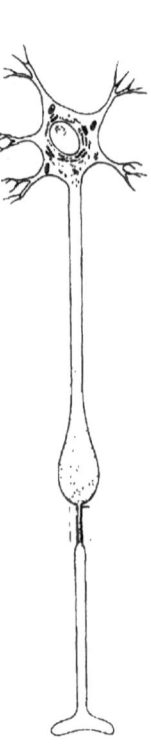

Axonal ballooning.

        (b)   Toxins (such as colchicine) that damage microtubules interfere with axoplasmic transport and result in shrinkage (atrophy) of distal axon

        (c)   Viruses (such as herpes simplex or herpes zoster) or toxins (such as tetanus toxin) enter axon terminals and are transported by retrograde transport to neuronal cell body

2.   **Schwann cell**

a.   Supporting cell for peripheral nerve axons; surrounded by basal lamina which bridges between adjacent Schwann cells along length of axon

b.   Myelinated fibers

 (1)   Schwann cell forms single internodal myelin sheath for one axon

 (2)   Myelin sheath results from fusion of layers of Schwann cell membrane to produce tight spiral lamellae around axon

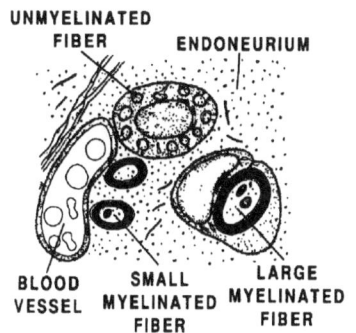

Structures of peripheral nerve fascicles.

  (a)   Inner (cytoplasmic) layers of plasma membrane fuse to form dense **period line**, while outer layers of plasma membrane fuse to form indistinct **interperiod line**

  (b)   External mesaxon — connection between myelin sheath and plasma membrane enclosing outer sleeve of Schwann cell cytoplasm

  (c)   Internal mesaxon — separation of inner myelin lamellae at interperiod line to connect with thin layer of periaxonal Schwann cell cytoplasm

 (3)   Schmidt-Lanterman cleft (incisure)

  (a)   Discontinuity in myelin sheath due to separation of period line

  (b)   Cytoplasmic connection between outer Schwann cell cytoplasmic sleeve and inner periaxonal Schwann cell cytoplasm

126

(c)   Involved in myelin maintenance, remodeling, and turnover

(4) Pi (π) granules of Reich

(a)   Schwann cell cytoplasmic granules that stain metachromatically and contain lamellar material

(b)   Increase with aging and with disorders causing axonal degeneration

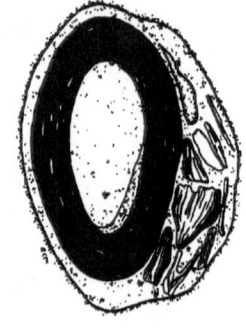

Pi granules in Schwann cell cytoplasm of myelinated fiber.

(5) **Node of Ranvier**

(a)   Region where myelin sheath ends and consecutive Schwann cells meet

(b)   Functional area for ion movement during saltatory conduction of nerve action potential

(c)   Cytoplasmic tongues of adjacent Schwann cells extend beyond termination of myelin sheath and interdigitate to cover nodal gap

(6) Direct relationship between axon diameter, myelin sheath thickness, and myelin segment length (distance between adjacent nodes of Ranvier or internodal length)

(a)   Outside diameter of myelinated fibers (axon plus myelin sheath) varies from 1 μm to 20 μm; largest fibers have myelin sheath thickness of 2 μm

(b)   Internodal lengths

i)   Internodal length increases during growth and development from embryonic length of 100 μm to adult internodal lengths which vary from 100 μm to 1800 μm (100 times fiber diameter)

ii) Following injury, new myelin sheath is of embryonic length, resulting in shorter remyelinated internodal segments

(7) Saltatory conduction

(a) Type of action potential propagation in myelinated fiber

(b) Depolarization "jumps" from one node of Ranvier to next

i) Local depolarization at one node takes about 15 μsec to excite next node (regardless of length of internodal segment)

ii) Largest diameter fibers with longest internodal segments conduct fastest

iii) Fiber conduction velocity (in meters per second) is approximately equal to 6 times fiber diameter (in micrometers)

c. Unmyelinated nerve fibers

(1) Contained within Schwann cell cytoplasm; single Schwann cell contains as many as ten unmyelinated fibers

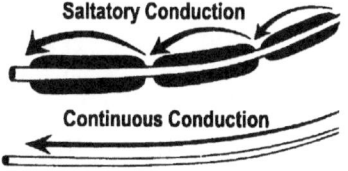

(2) Cytoplasmic processes of adjacent Schwann cells interdigitate with each other

(3) Continuous conduction — type of action potential conduction in which depolarization slowly travels along membrane

C. Membrane potentials

1. Nerve cell membrane is selectively permeable to ions with major intracellular ion being potassium and extracellular ion being sodium resulting in intracellular charge (resting potential) of about –70 mV compared to extracellular environment

    a.   Membrane ion pumps maintain charge by maintaining ion gradients; major ion pump is ATPase-dependent sodium-potassium pump that transports sodium ions out of cell in exchange for pumping potassium ions into cell

    b.   Hyperpolarization — change in transmembrane potential toward higher negative value

    c.   Depolarization — change in transmembrane potential toward less negative or positive value

2.   Action potential

    a.   Depolarization — reduction of local transmembrane potential below threshold level (about –50 mV) results in precipitous change in local membrane permeability with opening of voltage-dependent (gated) sodium ion channels allowing explosive passage of sodium ions such that transmembrane potential changes toward +50 mV

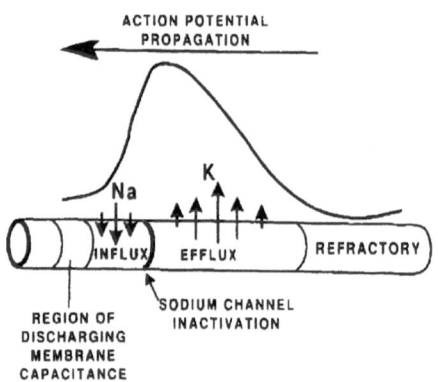

Action potential propagation along excitable membranes such as unmyelinated axons and muscle fiber membranes.

    b.   Repolarization — open sodium channels are quickly closed (inactivated), while voltage-dependent potassium ion channels open allowing efflux of potassium ions to balance earlier sodium ion influx, thereby repolarizing membrane toward resting potential

    c.   Current flow resulting from depolarization event triggers depolarization of adjacent areas of fiber, ultimately resulting in action potential that propagates over whole length of fiber

D.   Motor neurons

1.   Anterior horn cells of spinal cord or motor neurons in cranial nerve nuclei send motor axons to skeletal (striated) muscles

2.   Axons enter muscle belly at motor point and arborize into multiple terminal branches, each of which innervates single muscle fiber at its neuromuscular junction

3.   Motor unit

   a.   Composed of one anterior horn cell, its axon and terminal branches, and all muscle fibers innervated by that axon and its branches

   b.   Motor units are small for muscles involved with fine coordinated movements (hand and eye) and large for postural muscles:

      (1)   Quadriceps muscle — 3000 muscle fibers per motor unit

      (2)   Intrinsic hand muscles — 30 muscle fibers per motor unit

      (3)   Extraocular muscles — 3 muscle fibers per motor unit

E.   Sensory neurons — dorsal root ganglia or cranial nerve sensory ganglia

1.   Nerve cell bodies of primary sensory neurons for spinal nerve roots (dorsal root ganglia) or sensory roots of trigeminal nerve (semilunar ganglion and mesencephalic nucleus of trigeminal nerve), facial nerve (geniculate ganglion), and glossopharyngeal and vagus nerves (superior and inferior ganglia of vagus nerve)

2.   Perineuronal satellite cells — supporting Schwann cells that surround each spherical nerve cell body

3.   Unipolar neurons — single process that bifurcates to extend peripherally and centrally; large neurons have myelinated axons, while smaller neurons have unmyelinated axons

4.   Central processes branch to provide segmental, ascending, and descending connections, while peripheral processes connect to sensory receptors

F.   Specialized sensory receptors

1.   Free nerve endings — epidermal termination of fine myelinated and unmyelinated nerve fibers in dermis; sensitive to touch, pain, or temperature

2. Hair follicle endings — endings of small myelinated fibers that enwrap hair follicles; sensitive to touch

3. Merkel discs (touch corpuscles) — area of epidermal and dermal thickening containing large myelinated nerve fiber terminals in contact with layer of Merkel cells sending processes that indent overlying epidermal squamous cells; sensitive to touch

4. Ruffini corpuscles — complex network of dilated nerve endings in dermal papillae; sensitive to touch

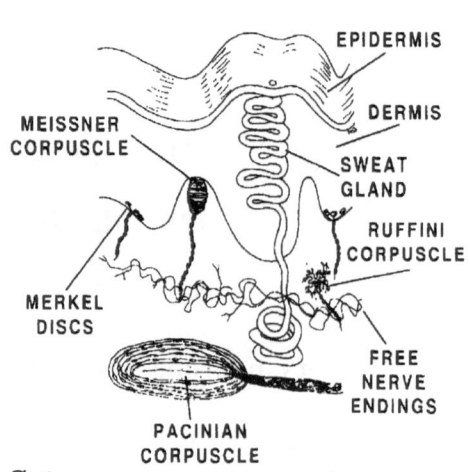

5. Krause end bulbs — connective tissue capsule surrounding numerous fine nerve endings of myelinated nerve fiber, mostly in mucous membranes; sensitive to temperature and pain

6. Meissner corpuscles — encapsulated nerve endings of large myelinated fibers, in palmar and plantar skin; involved in fine discrimination tactile (touch) sensation

Cutaneous sensory receptors.

7. Pacinian corpuscles — oval corpuscle composed of numerous concentric layers of cells and fluid surrounding nerve ending of large myelinated nerve fiber; found in subcutaneous tissue, fascia surrounding joints and tendons, and mesentery; sensitive to mechanical deformation from touch, pressure, vibration

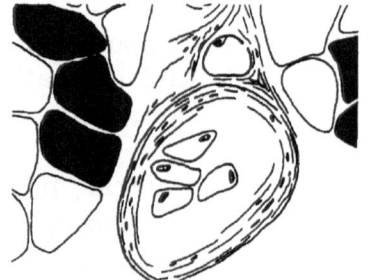

8. Golgi tendon organs — free nerve endings ramifying around collagen bundles in tendon; afferent nerve fiber silent when there is no tension on tendon, but active with tension on tendon from muscle stretch or muscle contraction

Cross section of muscle spindle containing four intrafusal muscle fibers.

9. Golgi-Mazzoni endings — resemble small Pacinian corpuscles, in soft tissue surrounding joints; sensitive to mechanical distortion from joint movement

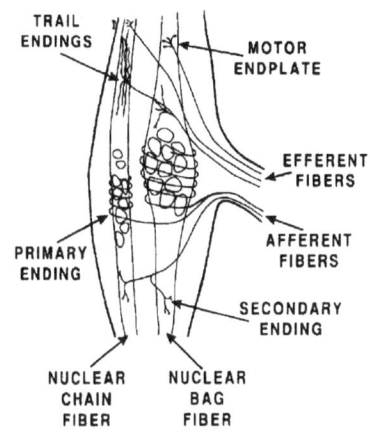

Muscle spindle.

10. Muscle spindle — third most complicated sensory organ behind eye and ear

    a. Capsule encloses specialized (intrafusal) muscle fibers, motor nerves, and sensory (afferent) nerves

    b. Intrafusal muscle fibers

        (1) Distinct from extrafusal skeletal muscle fibers

        (2) 3 to 15 muscle fibers per spindle

        (3) Nuclear bag fiber — nuclei clumped at center of fiber

        (4) Nuclear chain fiber — single row of nuclei

    c. Innervation

        (1) Primary (annulospiral) endings — large myelinated sensory axon endings that coil around center of each intrafusal muscle fiber

        (2) Secondary (flower-spray) endings — medium-sized myelinated sensory axon endings adjacent to primary endings on nuclear chain fibers

        (3) Motor innervation — γ-efferent fibers ending on intrafusal muscle fibers which are multiply innervated

            (a) Motor endplate — located in polar region of intrafusal fiber

            (b) Trail endings — small nerve terminal dilations spread over large area of muscle fiber in equatorial region

d. Ends of spindle are attached to connective tissue associated with extrafusal muscle fibers; intrafusal fibers are arranged in parallel with extrafusal fibers

(1) Muscle stretch increases tension on muscle spindle with resultant firing of spindle sensory (afferent) fibers

(2) Contraction of extrafusal muscle decreases tension on spindle unless intrafusal muscle also contracts

G. Neuromuscular junction

1. Area where nerve terminal comes into contact with specialized portion of muscle cell membrane (muscle endplate) that contains receptors sensitive to acetylcholine

2. Axonal action potential reaching nerve terminal, causes additional voltage-dependent calcium ion channels to open allowing calcium influx into nerve terminal; calcium mediates release of acetylcholine from synaptic vesicles into synaptic cleft between nerve terminal and muscle endplate; acetylcholine diffusing across synaptic cleft stimulates muscle acetylcholine receptors causing local depolarization; summation of local depolarizations exceeds threshold (endplate potential) and results in action potential propagation over muscle membrane with consequent muscle fiber contraction

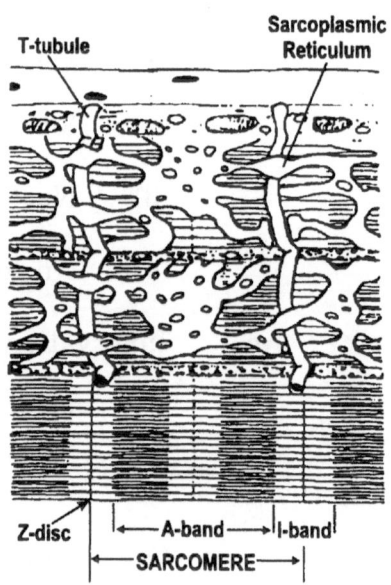

Muscle fiber structure.

H. Muscle fibers

1. Each skeletal muscle is composed of multiple striated muscle fibers enclosed by fascial connective tissue sheath

2. Muscle fibers are multinucleate cells surrounded by basal lamina, bounded by plasma membrane (sarcolemma), and containing thousands of cytoplasmic (sarcoplasmic) **myofibrils** composed of **actin** and **myosin** (contractile apparatus)

3. Each muscle fiber has single **acetylcholine receptor-rich endplate** opposite motor nerve terminal

4. Muscle fiber type

   a. Muscles are mosaics of muscle fibers (myofibers) distinguishable by morphologic, physiologic, biochemical properties into three types — type I, type IIA, and type IIB

   b. Muscle fiber type is not intrinsic property of muscle fiber, but is directed by properties of innervating axon (all muscle fibers in particular motor unit are of same fiber type)

   c. Muscle fiber type can change following denervation and subsequent reinnervation by different nerve

   ATPase-stained cross section of muscle showing mosaic of light type I fibers and dark type II fibers.

   d. Mosaic pattern for each individual is partially genetically determined and varies among different muscles and different individuals

5. Muscle fiber development

   a. During embryonic development, myoblasts (muscle cell precursors) align between tendons, fuse to form myotubes (multinucleate cell with row of central nuclei), and then differentiate to form mature muscle cells (fibers) containing contractile apparatus

   b. Satellite cells

      (1) Myoblast which did not fuse with other myoblasts in myotube formation, but remained separate from mature muscle fiber while still enclosed within same basal lamina

      (2) Following destruction of muscle fiber (from any insult), replacement muscle fiber is regenerated by satellite cells which divide to produce myoblasts that align in row (aided by basal lamina scaffolding), fuse to form myotube, and then mature into muscle fiber

    (3)   Unfused myoblasts remain as satellite cells to repeat regenerative process if necessary

  c. Myofiber size

    (1)   Myofibers increase in size from approximately 10 μm to 15 μm diameter at birth to approximately 50 μm to 60 μm diameter at puberty

    (2)   Increase in size of type II fibers is particularly influenced by activity and by androgenic stimulation

    (3)   Type II fibers tend to be larger in males than females

    (4)   Disuse (for example, as occurs with casting of arm or leg) results in atrophy (reduced size) of type II fibers

    (5)   Greatly increased activity (work hypertrophy) produces increase in size of type II fibers; this can be facilitated by androgenic steroids (particularly exogenously administered)

6. Muscle fiber contraction

  a. Endplate — single specialized region on each muscle fiber containing acetylcholine receptors; generates action potential (endplate potential) following acetylcholine release from nerve terminal

  b. Contractile apparatus

    (1)   Regularly arranged myofibrils producing striated appearance

    (2)   Functional unit of contraction is sarcomere which is area between adjacent Z-disks

    (3)   Actin (thin) filaments extend from each side of Z-disk

    (4)   Myosin (thick) filaments overlap with ends of actin filaments in center of sarcomere

  c. T-tubule system — network of tubules running transversely through muscle fiber at level of A-band/I-band junction with walls continuous with sarcolemma and lumen continuous with extracellular environment

d.   T-tubular triad — two terminal sacs of sarcoplasmic reticulum in close contact with T-tubules

e.   Sarcoplasmic reticulum — specialized endoplasmic reticulum of muscle fibers that sequesters calcium ions; calcium-ATPase pump removes free calcium ions from sarcoplasm

CHARACTERISTIC		MUSCLE FIBER TYPE		
		TYPE I	TYPE IIA	TYPE IIB
Color		Red	White	White
Fiber size		Small	Large (males > females)	Large (males > females)
ATPase histochemistry		Light	Dark	Dark
Cellular organelles	Mitochondria	Numerous	Moderate	Few
	Sarcoplasmic reticulum	Abundant	Minimal	Minimal
	Myoglobin	Large amount	Moderate amount	Small amount
Fuel stores	Glycogen	Low	High	High
	Triglycerides (lipid)	High	High	Low
Energy metabolism	Oxidative (aerobic)	High	High	Low
	Glycolytic (anaerobic)	Low	High	High
Phosphocreatine stores		High	Moderate	Low
Vascular supply	Surrounding capillaries	Numerous	Some	Few
	Blood flow	Good	Moderate	Rapidly insufficient
Twitch (contraction)	Speed	Slow	Fast	Fast
	Tension (force)	Low (weak)	High (strong)	High (strong)
	Fatigue	Resistant (Fatigues slowly)	Resistant (Fatigues slowly)	Sensitive (Fatigues quickly)
Functional role and type of contractions		Postural muscles; prolonged (sustained)	Voluntary; rapid, repeated or prolonged	Voluntary; rapid, strong, short duration ("quick burst")
Motor unit characteristics	Anterior horn cell size	Small	Large	Large
	Axon diameter	Small	Large	Large
	Axon conduction speed	Slow	Fast	Fast
	Motor unit size	Fewer fibers	Many fibers	Many fibers
	Recruitment threshold	Early	Late	Late
	Firing frequency	Low	High	High

f. Excitation-contraction coupling — action potential traveling over sarcolemma also travels down T-tubular system, where depolarization current transiently changes permeability of adjacent sarcoplasmic reticulum cistern resulting in release of calcium into sarcoplasm

g. Contraction — calcium ions diffusing through intermyofibrillar network interfere with troponin-tropomyosin inhibition of actin-myosin binding setting off chain of events in which ATP is cleaved to ADP, actin filaments slide along myosin filaments, and sarcomere shortens

h. Relaxation — sequestration of calcium ions from sarcoplasm permits troponin-tropomyosin inhibition of actin-myosin binding allowing actin filaments to slide back to resting position

II. Electrophysiologic studies

A. **Nerve conduction velocity (NCV)**

1. Recording of speed of propagation of electrical impulse by **largest myelinated axons** in nerve

2. **F-wave latency** evaluates conduction velocity in **proximal** nerve and nerve root

3. **H-reflex** evaluates **reflex arc** in lower extremity

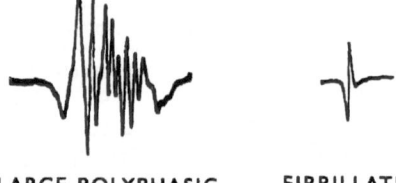

EMG motor unit potentials.

4. Abnormalities

a. Loss of nerve (axon) continuity leads to failure of impulse conduction

b. Demyelination of otherwise intact axons leads to slowing of conduction.

B. **Electromyography (EMG)**

1. Recording of electrical activity of muscle fibers using needle electrode inserted into muscle belly

2. Abnormalities

    a. **Motor unit potentials**

        (1) **Large amplitude polyphasic potential** — indicates reinnervation of denervated muscle fibers by axonal sprouting from adjacent intact nerves, producing enlarged motor units with increased numbers of muscle fibers per anterior horn cell

        (2) Decreased amplitude indicates reduced size of motor unit due either to partial denervation or destruction of muscle fibers

    b. **Fasciculations** — abnormal uncoordinated firing of muscle fibers in a single motor unit due to excessively irritable motor nerve; often **visible as muscle twitches**

    c. **Fibrillations** — twitchings of **single denervated muscle fibers** (visible by surgical exposure of muscle belly)

C. Sensory potentials

    1. Recording of computer averaged nerve action potentials generated by electrical stimulation of cutaneous nerves in digits

    2. Abnormalities — reduced amplitude or longer latency of nerve action potentials indicates neuropathy involving sensory fibers

D. Somatosensory evoked potentials

    1. Computer averaged potentials recorded over spine and head following stimulation of extremity nerve

    2. Abnormalities — reduced amplitude or longer latency of potentials indicates damage to central sensory pathways

III. Pathology

A. Definition of nerve abnormalities

    1. **Mononeuropathy — involvement of single nerve** due to local process (such as trauma, entrapment, vascular disease, or infection)

2. **Mononeuropathy multiplex** — asymmetric, **multiple nerve** abnormality (multiple mononeuropathy); usually due to **compromise of vasa nervorum** with consequent **nerve infarction** (such as in polyarteritis nodosum, diabetes mellitus, systemic lupus erythematosus, Wegner's granulomatosis, or rheumatoid arthritis)

3. **Polyneuropathy** — generalized **symmetrical** multiple nerve dysfunction, with **distal greater than proximal** involvement and longer nerves involved more than shorter nerves; due to metabolic derangement of nerve metabolism

4. **Radiculopathy** — involvement of spinal **nerve root**; isolated finding (such as in trauma or intervertebral disc protrusion) or part of polyneuropathy or mononeuritis multiplex

5. **Cranial neuropathy** — abnormality of **cranial nerve**

6. **Ganglionopathy** — abnormality of **nerve ganglion**; associated with sensory symptoms

7. **Plexopathy** — abnormality of **brachial or lumbar plexus**; idiopathic or result of trauma, local compression, tumor, infection or delayed effect of radiotherapy

8. **Autonomic neuropathy** — abnormality of **autonomic neurons** (with consequent anhidrosis, orthostatic hypotension, pupillary reflex paralysis, loss of lacrimation and salivation, impotence, and bowel and bladder dysfunction); most commonly occurs as a part of a **generalized polyneuropathy** (such as in diabetes mellitus), but also as an autosomal recessive disease (**Riley-Day syndrome**) or as part of a parkinsonian neurodegenerative disorder (**Shy-Drager syndrome**)

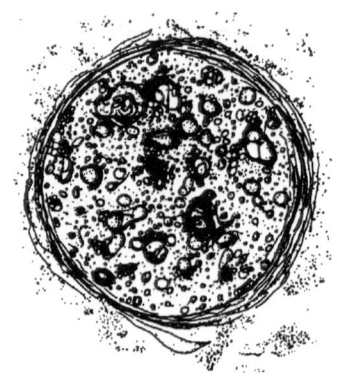

Cross-section of myelinated fiber undergoing wallerian degeneration.

9. **Motor neuropathy** — abnormality of **motor nerve fibers** either from direct damage to axons or due to death of anterior horn cells

10. **Sensory neuropathy** — abnormality of **sensory nerve fibers** from direct damage to axons or damage to neurons in sensory ganglia

11. **Sensorimotor neuropathy** — abnormality of both **sensory and motor nerve fibers**; most common form of neuropathy

B. Pathophysiology of nerve disease — basic responses by nerve to injury are similar regardless of underlying disease process

1. **Axonal degeneration**

   a. Degeneration of nerve axon from damage locally or in more proximal portion of nerve fiber

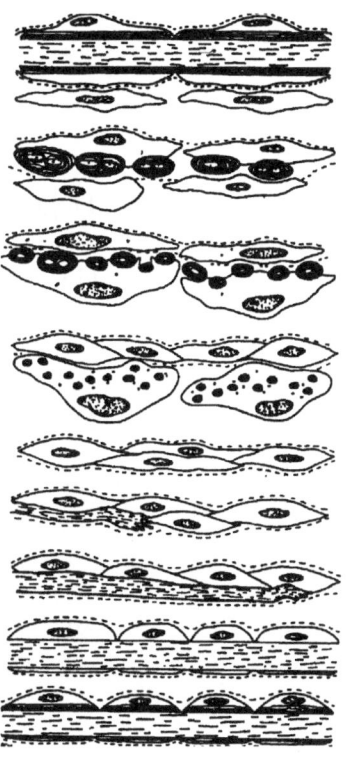

Pathologic changes of wallerian degeneration, axonal regeneration, and remyelination.

   b. **Wallerian degeneration** — change in nerve fibers occurring **distal to site of a focal destructive lesion** of axons or following destruction of neuronal cell body

      (1) Degenerating organelles accumulate at nodes, followed by fragmentation of axolemma, and then disintegration of axoplasm

      (2) Myelin sheath collapses, splits along interperiod line, and breaks up into myelin ovoids (lamellar myelin debris)

      (3) Phagocytosis by Schwann cells (myelin ovoids) and macrophages (axonal debris and myelin ovoids)

      (4) Proliferation of Schwann cells

      (5) Band of Büngner — interdigitation of processes of proliferating Schwann cells within residual basal

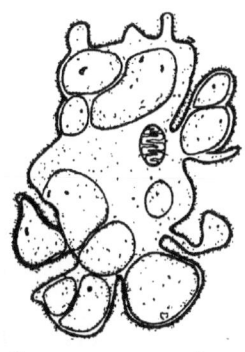

Cross section of band of Büngner.

lamina, with formation of Schwann cell "tubes" (composed of stacks or rows of proliferated Schwann cells within original surrounding "tube" of basal lamina), through which regrowth of axon from proximal stump is possible

c. Degeneration of unmyelinated fibers

   (1) Collagen pockets — following degeneration of unmyelinated axons, Schwann cells envelop longitudinal bands of collagen (in place of axons)

d. **Distal axonal degeneration**

   (1) Indolent process characterized by gradual distal-to-proximal (so-called **"dying back"**) axonal degeneration associated with secondary myelin destruction

   (2) Disturbance of axoplasmic transport deprives distal axon of essential organelles, proteins, and metabolites resulting in degeneration of distal parts of axon; can be due to abnormalities in nerve cell body or in proximal portions of axon

2. **Axonal regeneration**

   a. Central chromatolysis

      (1) Regenerative change in nerve cell body following axonal injury associated with increased metabolic activity

      (2) Enlargement of nerve cell body, eccentric displacement of nucleus, dispersal of Nissl substance (rough endoplasmic reticulum and polyribosomes) to periphery of cell, and increased RNA and protein synthesis

   b. Initial ballooning of proximal intact axon stump (growth cone) caused by accumulation of organelles, neurofilaments, and neurotubules transported from nerve cell body

   c. Extension from growth cone of multiple neurites (axon sprouts), which grow into bands of Büngner (Schwann cell "tubes") and become remyelinated

d. Regenerative cluster — often several axon sprouts enter same band of Büngner, resulting in cluster of myelinated fibers enclosed within single basal lamina

e. **Neuroma**

   (1) Tangled mass of axonal sprouts resulting from aberrant axonal regeneration with sprouts unable to find bands of Büngner (Schwann cell "tubes") in which to regrow

   (2) Morton's neuroma — painful neuroma involving plantar interdigital nerve; common in elderly individuals

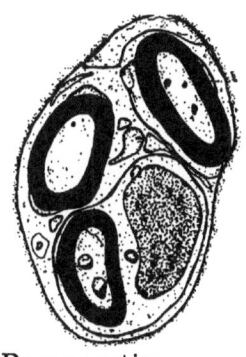

Regenerative cluster of myelinated fibers in single Schwann cell basal lamina.

3. Segmental demyelination (primary myelin degeneration)

a. Destruction of individual myelin internodes or Schwann cells, while **axonal integrity is preserved**

b. Primary segmental demyelination — metabolic disturbance in Schwann cell results in myelin sheath breakdown

   (1) Causes include lead intoxication, diphtheritic neuropathy, and minimal ischemia (Schwann cells are more sensitive to ischemia than axons)

   (2) Initial widening of nodal gap (retraction of myelin at nodes of Ranvier) followed by breakdown of myelin and phagocytosis of myelin ovoids by Schwann cells and macrophages

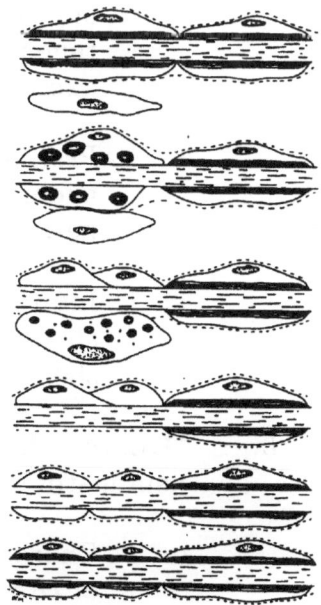

Pathologic changes of segmental demyelination and remyelination.

c. Autoimmune demyelination — immunologic attack on peripheral nerve myelin sheaths

(1) Occurs in Guillain-Barré syndrome and chronic inflammatory demyelinating polyradiculoneuropathy

(2) Macrophages invade Schwann cell basal lamina and penetrate myelin sheath at interperiod line, separating and unrolling myelin lamellae; this is followed by breakdown of myelin and phagocytosis of myelin ovoids by macrophages

Cross-section of demyelination due to immune-mediated macrophage delamination and digestion of myelin sheath.

4. Myelin regeneration (**remyelination**)

a. Follows either primary demyelination (over previously demyelinated segments of intact axons) or after axonal regeneration (over naked regenerating axons)

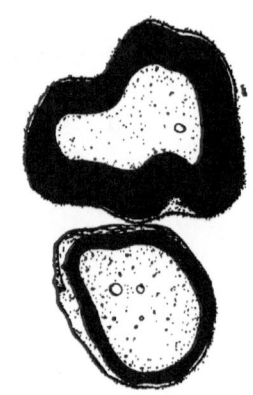

b. Proliferating Schwann cells align along axon; surviving Schwann cells produce new myelin sheath

c. Internodal length of each new myelin sheath segment is shorter and myelin sheath is thinner (fewer lamellae)

Cross-sections of axons of similar diameter: remyelin-ated fiber has thinner myelin sheath.

5. **"Onion bulb" formation** (hypertrophic neuropathy)

a. Characteristic morphologic change in myelinated fibers in certain **chronic demyelinating or hereditary neuropathies**, including chronic inflammatory demyelinating polyradiculoneuropathy, Charcot-Marie-Tooth disease, and Déjérine-Sottas disease

b. Repeated cycles of demyelination and remyelination result in florid proliferation of Schwann cells which form **concentric layers** around axon (resembling onion bulbs in cross section)

6. **Nerve trauma**

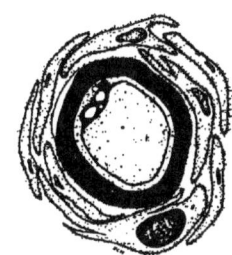

Onion-bulb formation.

a. **Neurapraxia** — nerve is injured (i.e. compression) **without physical axonal disruption**, leading only to transitory "physiologic" dysfunction

b. **Axonotmesis — axon is disrupted** (i.e. crushed) but Schwann cell tubes remain to facilitate axonal regeneration and substantial recovery

c. **Neurotmesis — completely severed nerve** such that amount of recovery depends on adequacy of alignment of fascicles in reanastomosis

C. Selected specific neuropathies

1. **Guillain-Barré Syndrome** — autoimmune demyelinating neuropathy

a. Clinical symptoms of progressive **symmetric motor weakness**, usually beginning distally in legs and progressing proximally and cephalad

b. Elevated CSF protein with few cells (**albuminocytologic dissociation**)

c. Prolonged F-wave latency, slowing of nerve conduction velocities, and absent H-reflex.

2. **Diabetic neuropathy** — common cause of neuropathy, resulting in polyneuropathy (metabolic) or mononeuritis multiplex (ischemic)

a. Symmetric sensorimotor polyneuropathy

(1) Distal axonal degeneration resulting in **loss of sensation in feet** and to lesser extent hands ("stocking-glove distribution")

       (2)   Altered nerve metabolism results from marked thickening of basement membranes which surround endoneurial blood vessels, Schwann cells, and perineurial cells, which impedes transport of essential nutrients; elevated endoneurial glucose levels interfere with normal metabolic pathways

       (3)   **Autonomic neuropathy** — involvement of autonomic nerves resulting in decreased sweating, postural hypotension, impotence, **bladder dysfunction** (usually atonic bladder), and **disturbed gastrointestinal motility**

   b.   Mononeuritis (or mononeuritis multiplex) — nerve ischemic infarction due to occlusion of vasa nervorum (nerve vasculature)

3.   Polyarteritis nodosa

   a.   Prototypic ischemic neuropathy due to necrotizing vasculitis destroying vasa nervorum

   b.   Produces ischemic infarction in various nerves and portions of nerves, resulting in clinical syndrome of mononeuritis multiplex

4.   **Charcot-Marie-Tooth disease (peroneal muscular atrophy syndrome**, hereditary motor and sensory neuropathy type I)

   a.   Clinical symptoms of slowly progressive **distal muscle atrophy, pes cavus**, atrophy of lower legs ("stork legs" or "inverted champagne bottle legs"), and sensory loss in feet and hands

   b.   **Autosomal dominant** inheritance

   c.   Axonal loss and onion bulb formation in peripheral nerves

5.   Amyloid neuropathy

   a.   Neuropathy associated with heterogeneous group of disorders characterized by deposition of amyloid (highly insoluble fibrillar protein arranged as **beta-pleated sheet**)

       (1)   Familial amyloid polyneuropathies — amyloid protein produced from mutated transthyretin (prealbumin) protein

(2)   Immunoglobulin-derived amyloid polyneuropathy — altered immunoglobulin light chain protein produces amyloid deposits (AL amyloid)

(3)   Secondary amyloidosis — chronic inflammatory disorders result in deposition of amyloid (AA amyloid)

b. Amyloid deposits found in epineurium, around endoneurial blood vessels, as nodular masses in endoneurium, and in connective tissue surrounding nerve

c. Mechanism of nerve damage not well understood, except in cases of compressive damage from amyloid deposits (as in flexor retinaculum of wrist compressing median nerve to produce carpal tunnel syndrome)

6. **Herpes zoster (shingles)**

a. During initial varicella (chickenpox) infection, virus enters sensory nerve terminals and by retrograde axoplasmic transport is carried to nerve cell bodies in **sensory ganglia** where it becomes **latent**

b. **Virus reactivation** follows illnesses that alter immune status

(1)   Viral particles are transported by anterograde axoplasmic transport to axon terminals resulting in skin infection **(vesicular rash)** in **dermatomal distribution** of involved nerve; most often involves thoracic or trigeminal ganglia

(2)   Hemorrhagic ganglionitis — involved sensory ganglion has necrotic neurons and satellite cells containing **Cowdry type A (viral) inclusions,** along with hemorrhage and marked inflammatory cell infiltration

(3)   Healing associated with fibrosis along with loss of ganglion cells and axons

7. **Leprosy** — common neuropathy in underdeveloped countries due to **invasion of nerves** by acid-fast bacillus *Mycobacterium leprae* producing palpable nerve enlargement

a. **Tuberculoid leprosy** — granulomas involving skin with destruction of local nerves

b. **Lepromatous leprosy**

    (1) Hematogenous spread of organism resulting in **diffuse** infiltration of skin and peripheral nerves with widespread intracellular accumulation of innumerable organisms, particularly in unmyelinated axons, Schwann cells of unmyelinated axons, and endothelial cells

    (2) Progressive destruction of axons and Schwann cells

    (3) Greatest involvement of **cooler parts of body** such as pinnae of ears, tip of nose, and dorsum of feet and hands

8. **Motor neuron disease**

a. **Amyotrophic lateral sclerosis** (ALS)

    (1) Clinical symptoms

        (a) **Progressive generalized lower motor neuron signs of muscle weakness** and **wasting (atrophy)** with **fasciculations**

        (b) **Upper motor neuron** signs of spasticity with hyperactive tendon reflexes and extensor plantar responses (Babinski reflexes)

        (c) No significant sensory abnormality

    (2) Pathologic features

        (a) **Gross atrophy of precentral (motor) gyri and anterior spinal roots**

        (b) **Corticospinal tract degeneration** — loss of axons and myelin accompanied by reactive gliosis in spinal cord anterior and lateral columns and medullary pyramids

        (c) **Lower motor neuron degeneration**

            i) Marked loss of motor neurons accompanied by reactive gliosis of spinal cord anterior horns and medullary hypoglossal nuclei

ii)   Loss of axons in anterior spinal roots

iii)   **Denervation changes in skeletal muscle**

b. **Werdnig-Hoffmann disease (infantile spinal muscular atrophy** type I)

   (1) **Floppy infant** with progressive weakness, tongue fasciculations, feeding difficulties, and death from ventilatory insufficiency before age 2 years

   (2) **Autosomal recessive** disorder

   (3) Pathologic features

       (a) Marked **loss of motor neurons** from spinal cord anterior horns and medullary hypoglossal nuclei, along with reactive gliosis; remnant motor neurons shrunken; neuronophagia (phagocytes removing necrotic neuron) evident

       (b) No abnormality of corticospinal tracts

       (c) Anterior spinal roots markedly atrophic and often contain bundles of astrocytic processes in proximal segments

       (d) Muscle atrophy with characteristic massively hypertrophic type I muscle fibers among numerous atrophic muscle fibers

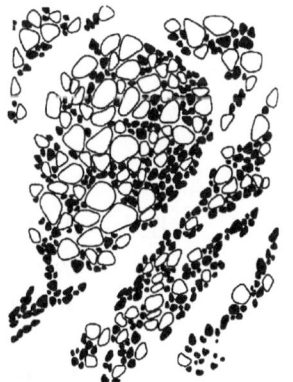

ATPase stain shows type I fiber hypertrophy in Werdnig-Hoffmann disease.

c. **Kugelberg-Welander disease** (juvenile spinal muscular atrophy)

   (1) Clinical symptoms of **progressive proximal weakness** and fasciculations beginning in childhood or adolescence

   (2) **Autosomal recessive** or **autosomal dominant** inheritance

  (3) Pathologic features — **loss of motor neurons** and accompanying reactive gliosis in spinal cord anterior horns and medullary hypoglossal nuclei; chronic neurogenic changes in muscle including atrophic fibers and fiber type grouping

D. Neuromuscular junction diseases

1. Myasthenia gravis

 a. Clinical symptoms of progressive **weakness after exercise ("fatiguable weakness")** with most severe involvement of **ocular muscles**, followed by **bulbar** and respiratory muscles, while extremity muscles are least affected

  (1) Simulates effect of **partial curarization** (exposure to low doses of nondepolarizing muscle relaxant *d*-tubocurarine)

  (2) Weakness can be exacerbated by drugs interfering with neuromuscular transmission, including succinylcholine, quinidine, procainamide, quinine (tonic water), penicillamine, streptomycin, kanamycin, polymyxin, lincomycin, tetracycline, gentamicin; safe drugs include penicillin, cephalothin, rifampin, vancomycin, amphotericin, nystatin

 b. Jolly test — progressive **decrement** in amplitude of compound muscle action potential recorded during 3 Hz repetitive nerve stimulation

 c. Frequently associated with **thymic abnormalities**, particularly **hyperplasia** in young women and **thymoma** in old men

 d. **Autoimmune** attack on muscle endplate acetylcholine receptors, with serum **anti-acetylcholine receptor (anti-AChR) antibodies (IgG)** detectable in most patients; other simultaneous autoimmune diseases common

Progressive decrement in amplitude of compound muscle action potential with 3 Hz stimulation in myasthenia gravis.

e. Muscle endplates show simplification of junctional folds, due to excessively rapid turnover due to antibody attack

2. **Lambert-Eaton (myasthenic) syndrome**

a. Clinical symptoms of progressive **generalized muscle weakness** (proximal greater than distal) that **improves with exercise**; absent tendon reflexes and mild sensory loss also evident

b. **Autoimmune** attack on presynaptic nicotinic motor nerve axon terminals, with serum IgG antibodies detectable in most patients

c. Commonly associated with **carcinoma**, particularly **small cell (oat cell) lung carcinoma**, or with other autoimmune diseases

3. **Botulism**

a. Clinical symptoms of acute, rapidly progressive **paralysis of extraocular and bulbar** (pharyngeal) muscles; skeletal muscle weakness and respiratory compromise follow within 2 to 3 days; **constipation** occurs from smooth muscle paralysis

b. Caused by ingestion of **exotoxin** (which blocks acetylcholine release from nerve terminals) produced by anaerobic gram-positive spore-forming rod bacteria *Clostridium botulinum*; most often follows eating improperly canned vegetables

c. Diagnosis confirmed by assay of toxin in serum or tainted food

d. **Neonatal botulism** occurs in infants whose gastrointestinal tracts are colonized by *Clostridium botulinum* with consequent absorption of exotoxin

E. Pathophysiology of muscle disease

1. **Necrosis and regeneration**

a. Focal muscle fiber injury — most of sarcolemma and cytoplasmic contents remain intact, allowing sarcolemmal membrane to reseal; increased RNA and protein synthesis replaces damaged or lost constituents

    b.   Extensive muscle fiber injury

        (1)   Final common pathway to muscle fiber death is accumulation of free sarcoplasmic calcium ions due to physical damage to sarcolemma or to depletion of energy stores

        (2)   Fiber death is followed by macrophage infiltration, phagocytosis, and release of muscle fiber contents into circulation; serum level of muscle enzyme creatine kinase (CK) is sensitive indicator of muscle fiber damage

    c.   If basal lamina remains intact, satellite cell proliferation along this scaffolding can regenerate new muscle fiber; minor damage to basal lamina can be repaired during satellite cell proliferation, but extensive damage to basal lamina often prevents fiber regeneration and results in fibrosis

2.   Muscle fiber hypertrophy and splitting

    a.   Fiber enlargement occurs normally with growth, but is aided by activity and by androgenic hormone influences

    b.   Fiber splitting to form two smaller fibers follows marked enlargement of fiber diameter

3.   Muscle fiber atrophy

    a.   **Denervation** (loss of innervation)

        (1)   Fiber becomes small and angulated; seemingly compressed by surrounding innervated fibers

        (2)   Receptors for acetylcholine appear over whole sarcolemma (**extra-junctional acetylcholine receptor**) making fiber sensitive to spontaneously generated depolarizations (**fibrillations**)

        (3)   If reinnervation does not occur, myofibrils are lost and remnant sarcolemma and nuclei ("pyknotic nuclear clumps") undergo phagocytosis by macrophages

    b.   **Type II fiber atrophy** — selective atrophy of type II muscle fibers produced by **disuse** (lack of activity), **endocrine disorders**

151

(including disturbances in thyroid or parathyroid hormone levels or **excess corticosteroids**), and as remote effect of cancer (carcinomatous myopathy)

c. **Type I fiber atrophy** — selective atrophy of type I fibers is characteristic of **myotonic dystrophy** and two congenital myopathies (centronuclear myopathy and congenital fiber type disproportion)

4. Target fibers — abnormality in fiber center consisting of three concentric alternating light and dark zones (resembling rifle target), usually associated with chronic neurogenic disease

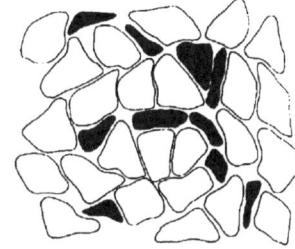

ATPase stain is used to reveal selective type II muscle fiber atrophy.

5. **Type grouping**

a. Reinnervation of denervated muscle fiber by axonal sprout from adjacent intact nerve axon enlarges motor unit of reinnervating axon

b. Reinnervated muscle fiber changes to fiber type of new innervating axon

c. Denervation and reinnervation of many fibers in area of muscle alters normal mosaic pattern (produced by intermingling of axons from many different motor units) to uniform fiber type pattern

d. **Group atrophy** — damage to reinnervating axon results in atrophy of whole group (type group) of fibers

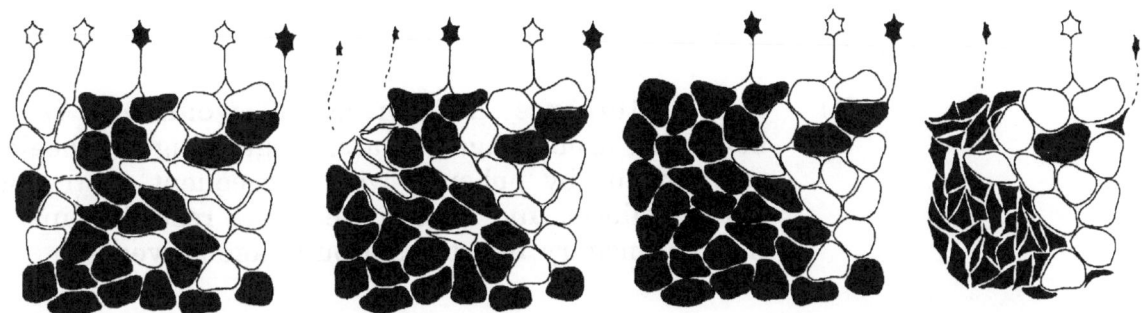

Axon damage causes muscle fiber atrophy, followed by reinnervation from axon sprouts producing fiber type grouping. Later axonal loss results in muscle fiber group atrophy.

F.   Selected specific muscle diseases

1.   **Polymyositis**

a.   Clinical symptoms — subacute progressive **proximal muscle weakness**, muscle discomfort or pain, and in some cases **skin involvement (dermatomyositis)** including purple discoloration of eyelids, skin erythema (dorsal surfaces of hands and over finger, elbow, or knee joints), and subcutaneous calcifications; elevated levels of serum creatine kinase

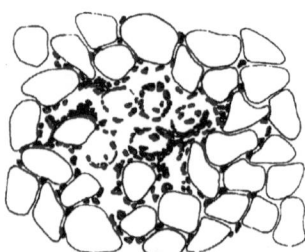

Chronic inflammatory cell infiltration in polymyositis.

b.   **Autoimmune** attack on skeletal muscle fibers; linked to occult cancers and to other autoimmune disorders

c.   Pathologic features

(1)   Diffuse infiltration of muscle interstitial connective tissue with lymphocytes, plasma cells, and macrophages

(2)   Muscle fiber necrosis and phagocytosis by macrophages (myophagocytosis)

(3)   Regeneration of some muscle fibers, while others are replaced by fibrosis

(4)   Calcification (rarely) in connective tissue surrounding muscle fascicles

2.   **Duchenne muscular dystrophy**

a.   Clinical symptoms

(1)   Gait disturbance and **proximal weakness** in boys beginning around age 3 years, progressing to wheelchair confinement by puberty, with subsequent contractures, kyphoscoliosis, and further weakness resulting in death from pulmonary infections around age 20 years

       (2)   Early clinical examination reveals enlarged calf muscles **(pseudohypertrophy)** which have "rubbery" or "woody" consistency

       (3)   Levels of serum creatine kinase greatly elevated (often 100 times normal)

  b.   Genetic defect due to **complete absence of muscle membrane protein dystrophin** normally produced by X-chromosome gene (Xp21 gene locus)

       (1)   Dystrophin gene locus — largest gene locus so far described; large size accounts for high mutation rate

       (2)   **Becker muscular dystrophy — partial deficiency** of dystrophin results in milder phenotype

       (3)   Females carriers of abnormal gene can manifest muscle symptoms due to abnormal **lyonization** (inactivation) of normal X-chromosome

  c.   Pathologic features

       (1)   Numerous necrotic fibers (undergoing phagocytosis) and regenerating fibers, which explains markedly elevated serum creatine kinase levels

       (2)   Marked endomysial fibrosis (between muscle fibers), which explains clinical muscle enlargement and abnormal consistency

3.   Facioscapulohumeral (FSH) dystrophy

  a.   Clinical symptoms

       (1)   Slowly progressive weakness of facial, shoulder, upper arm muscles, and lower leg muscles

       (2)   Characteristic inability to whistle or blow out candles, markedly atrophic pectoralis muscles, and atrophy of upper arm muscles with normal forearm muscles ("Popeye-the-sailorman" appearance)

  b.   Autosomal dominant pattern of inheritance

c. Pathologic features — nonspecific abnormalities of variability in muscle fiber size, patchy uneven muscle fiber staining (due to focal disruption of myofibrils), and occasional necrotic or denervated muscle fibers

4. Limb-girdle dystrophy

   a. Heterogeneous syndrome characterized by slowly progressive muscle weakness (greater proximally than distally) and elevation of serum creatine kinase levels

   b. Nonspecific pathologic features of variability of muscle fiber size and occasional degenerating and regenerating muscle fibers

5. **Myotonic muscular dystrophy**

   a. Multisystem disorder characterized by myopathy, **cardiac arrythmia**, **cataracts**, frontal baldness, gastrointestinal disturbances, endocrine dysfunction, gonadal atrophy, and intellectual and behavioral disturbances; widely varying spectrum of severity and age of clinical onset of symptoms

   b. Muscle abnormalities

      (1) Distal and facial weakness

      (2) **Myotonia** — muscle membrane abnormality that results in repetitive muscle fiber depolarizations following single voluntary contraction; clinically evident as muscular stiffness, while electromyography (EMG) reveals characteristic **"dive-bomber" potentials**

   c. Autosomal dominant disorder due to genetic defect on chromosome 19

   d. Pathologic features

      (1) Type I fiber atrophy

      (2) Numerous muscle fibers with disrupted myofibrils

      (3) Increased number of intrafusal muscle fibers (up to 60 fibers in some spindles)

6. **Mitochondrial myopathy**

a. Heterogeneous group of disorders involving various organ systems with **abnormal mitochondrial function**

b. Mutations observed in nuclear genes encoding mitochondrial proteins (mendelian inheritance pattern) or mitochondrial genes (maternal inheritance pattern)

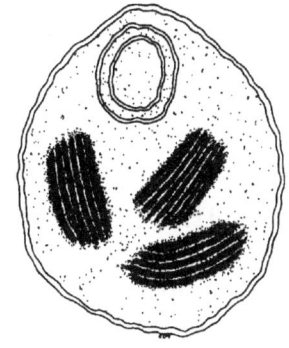

Mitochondrial para-crystalline inclusions.

c. Selected defined syndromes

    (1) **Leber's optic atrophy** — progressive optic atrophy

    (2) **MERRF** (myoclonus epilepsy with ragged red fibers) — myoclonus, seizures, ataxia, dementia, weakness, short stature

    (3) **MELAS** (mitochondrial encephalomyopathy, lactic acidosis, and stroke-like episodes) — headaches, seizures, elevated scrum lactate levels, stroke episodes

    (4) **Kearns-Sayre syndrome** — progressive external ophthalmoplegia, pigmentary retinopathy, cardiac conduction defects

d. Identifiable in muscle biopsies histologically as accumulated bright red subsarcolemmal staining in frozen sections and by electron microscopy as increased numbers of structurally abnormal mitochondria which often contain **paracrystalline inclusions**

SUGGESTED ADDITIONAL READING

Brooke MH: *A Clinician's View of Neuromuscular Diseases.* ed 2. Baltimore, Williams & Wilkins Co, 1986.

Brumback RA, Leech RW: *Color Atlas of Muscle Histochemistry.* Littleton, Mass, PSG Publishing Co, 1984.

Dubowitz V: *Muscle Disorders in Childhood.* London, W. B. Saunders Co, 1978.

Dyck PJ, Thomas PK, Lambert EH, et al (eds): *Peripheral Neuropathy*, ed 2. Philadelphia, W. B. Saunders Co, 1984.

Engel AG, Banker BQ (eds): *Myology*. New York, McGraw-Hill Book Co, 1986.

Ropper AH, Wijdicks EFM, Truax BT: *Guillain-Barré Syndrome*. Philadelphia, F. A. Davis Co, 1991.

Schaumburg HH, Berger AR, Thomas PK: *Disorders of Peripheral Nerves*. ed 2. Philadelphia, F.A. Davis Co, 1992.

Vinken PJ, Bruyn GW, Ringel SP (eds): *Diseases of Muscle-Part I. Handbook of Clinical Neurology. Volume 40*. Amsterdam, North Holland Publishing Co, 1979.

Vinken PJ, Bruyn GW, Ringel SP (eds): *Diseases of Muscle-Part II. Handbook of Clinical Neurology. Volume 41*. Amsterdam, North Holland Publishing Co., 1979.

Walton J: *Disorders of Voluntary Muscle*. ed 4. New York, Churchill Livingstone, 1981.

CHAPTER 10: PATHOLOGY OF TOXIC AND NUTRITIONAL DISEASES

I. **Central pontine myelinolysis**

    A. Clinical signs of progressive spastic quadriparesis and lower cranial nerve palsies often obscured by stupor or coma from associated systemic disease

    B. Pathologic features

        1. **Triangular or diamond-shaped, discolored, demyelinated** (and sometimes centrally cavitated) **area with relative preservation of axons and pontine neurons in basis pontis**; descending corticospinal tracts tend to be spared

        2. Depending on severity, lesion centered in midpons can extend to involve:

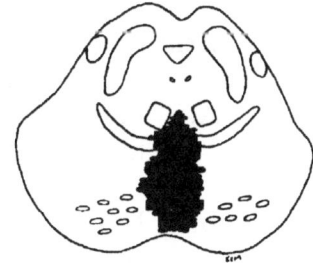

           a. Pontine tegmentum

           b. Cerebellar white matter

           c. Midbrain, thalamus, internal and external capsule, caudate, putamen

Central pontine myelinolysis.

    C. Originally attributed to chronic alcoholism, but now known to result from **too rapid correction (or overcorrection) of hyponatremia or serum hypo-osmolality**

II. **Marchiafava-Bignami disease**

    A. **Discolored, demyelinated** (and sometimes centrally cavitated) **area with relative preservation of axons in anterior central (midline) portion of corpus callosum**; lesions sometimes also identifiable in other central white matter tracts such as anterior commissure and optic chiasm

B. Clinical symptoms usually not evident, but lesions can be visualized with computed tomographic (CT) scan or magnetic resonance imaging (MRI)

C. Originally described in **Italian alcoholics consuming cheap red wine**, but subsequently identified in various other populations; pathogenesis is unknown

III. Methanol (methyl alcohol; wood alcohol) poisoning

A. Ingestion of as little as 60 mL can be fatal; commonly found as **adulterant in alcoholic beverages** (particularly illegally-produced "moonshine")

B. Initial symptoms of drunkenness, headache, abdominal pain, and visual loss, evolve into delirium and coma

1. Metabolism of methanol by hepatic alcohol dehydrogenase into formaldehyde and formic acid results in **severe metabolic acidosis**

2. **Formic acid disrupts axoplasmic flow in optic nerve**, resulting in **optic disc swelling** and axonal destruction

3. Following recovery, **blindness** persists due to retinal and optic nerve damage

IV. Ethylene glycol poisoning

A. Major constituent of automobile **antifreeze**; sometimes accidentally consumed by alcoholics; ingestion of 120 mL can be fatal

B. Initial presentation of drunkenness, followed by **generalized convulsions** and coma

C. Metabolic conversion of ethylene glycol by hepatic alcohol dehydrogenase into glycolic acid and oxalic acid produces severe **acidosis** and **oxalate crystal deposition** in **kidneys** (resulting in **uremia**) and in **brain** (resulting in **chemical meningitis** with increased numbers of CSF lymphocytes)

V. Lead poisoning

A. **Acute encephalopathy**

1. Infants or young children with **pica** (eating of non-nutritive substances such as dirt, clay, or flaking paint containing lead) present with **irritability, anorexia, lethargy, ataxia, seizures, and coma**

    2.   Often fatal due to **massive cerebral edema** and diffuse neuronal necrosis

    3.   Survivors have residual mental retardation, seizures, ataxia, and spasticity

  B. **Motor neuropathy**

    1.   Chronic lead exposure in adults produces pure motor neuropathy affecting most exercised muscles (hence, **wrist drop** or **foot drop**)

    2.   Toxic effect of lead on Schwann cells and myelin membranes results in segmental demyelination

  C. Associated with **anemia, basophilic stippling** of red blood cells, **"lead line"** along gingiva, **constipation**, and **colicky abdominal pain**

VI. **Arsenic poisoning**

  A. **Neuropathy — painful sensorimotor neuropathy** (red burning hands and feet); pathologic features of axonal degeneration affecting large fibers more than small fibers

  B. **Encephalopathy** — progressive fatigue, lethargy, headache, confusion, seizures, coma, and death; autopsy reveals punctate **white matter hemorrhages** ("brain purpura")

  C. Associated with **anemia, brown skin discoloration**, abdominal pain, plantar and palmar **hyperkeratosis**, and white transverse bands in nails (**Mees' lines**); increased levels of arsenic can be detected in hair and urine

VII. **Mercury poisoning** (organic or inorganic mercury)

  A. Adults — personality disturbance and **dementia** (use of mercury in felt hat production resulted in descriptive phrase: "mad as a hatter"), along with cerebellar ataxia, intention tremor, and **motor neuropathy**

  B. Children — **acrodynia** (Pink disease): swollen, red, cold, moist hands and feet, associated with irritability, insomnia, and anorexia

  C. Organic mercury can cross placenta causing mental retardation or cerebral palsy

    D.   Pathologic features (from studies of outbreaks of organic mercury poisoning)

        1.   Cerebral atrophy with patchy neuronal loss, greatest in calcarine cortex and precentral and postcentral gyri

        2.   Cerebellar atrophy with diffuse loss of Purkinje cells and granule cells

VIII. Thallium poisoning

    A.   Initial gastrointestinal symptoms followed by painful peripheral and autonomic neuropathy, behavioral disturbances, and seizures; alopecia and damage to sweat and sebaceous glands result in dry scaly skin

    B.   Pathologic features

        1.   Cerebral edema, patchy neuronal loss, and degeneration of dorsal columns

        2.   Peripheral nerve distal axonal degeneration ("dying-back" neuropathy)

IX. Carbon monoxide poisoning

    A.   **Hypoxia without cyanosis**

        1.   Carbon monoxide **displaces oxygen from hemoglobin** to form **carboxyhemoglobin,** with resultant **tissue hypoxia**

        2.   Initial symptoms of headaches and dizziness occur with 20% carboxyhemoglobin concentration, progress to lethargy with 40% concentration, coma and seizures with >50% concentration, and death from brain stem and myocardial dysfunction with >70% concentration

        3.   Individuals dying acutely have **cerebral edema** and **abnormal bright red color** of brain (and other tissues) due to presence of carboxyhemoglobin (**"cherry-red" appearance**)

    B.   Treatment by early removal from carbon monoxide source and administration of 100% oxygen can result in survival

        1.   Many survivors have no evidence of neurologic sequellae

        2.   Neurologic damage in survivors can occur in several patterns (separately or in combination):

        a.   **Pallidal necrosis** — bilateral necrosis of globus pallidus (inner segment greater than outer segment), due to hypoxia in vascular distribution of pallidal branches of anterior choroidal arteries

        b.   **Laminar ("pseudolaminar") cortical necrosis** — extensive, nearly continuous (but of variable thickness) band of cerebral cortical destruction associated with slit-like cavitation parallel to pial surface

        c.   **White matter destruction (leukoencephalopathy)** — patchy or confluent demyelination with relative preservation of axons; often associated with biphasic clinical course of recovery from acute symptoms followed by later deterioration

X.   **Wernicke-Korsakoff syndrome** — related to thiamine deficiency and most commonly observed among chronic alcoholics

    A.   **Wernicke disease (encephalopathy)** — abrupt onset of **nystagmus** and **lateral rectus palsy** ("cross eyes") progressing to **external ophthalmoplegia** (paralysis of eye movement) associated with truncal and gait **ataxia** and **global confusional state**

    B.   **Korsakoff's psychosis — chronic amnesic syndrome**; follows resolution of Wernicke disease symptoms of ataxia and external ophthalmoplegia

        1.   **Anterograde amnesia — inability to form new memories** despite relatively intact immediate recall and relatively preserved remote memory

        2.   **Confabulation** — tendency to **falsify memories** by **"filling in" gaps in memory** with information that sounds plausible but has little basis in reality; alertness, attentiveness, and other behavioral functions are normal

    C.   Pathologic features — vary with stage and severity of disease

        1.   **Wernicke disease** — symptoms result from **petechial hemorrhages, reactive edema, astrocytic proliferation (gliosis), and demyelination** (with relative preservation of neurons, although some neurons show central chromatolysis) involving various brain stem and diencephalic gray matter areas:

       a.   **Confusional state** — involvement of **mammillary bodies, hypothalamus,** and **periventricular thalamus** (particularly dorsomedial and anterior medial nuclei)

       b.   **Ophthalmoplegia** and **ataxia** — involvement of **periaqueductal gray matter** and **gray matter beneath floor of fourth ventricle**

   2.   **Korsakoff's psychosis**

       a.   Bilaterally symmetric **severe neuronal loss and gliosis** of **dorsomedial nuclei of thalamus** and mammillary bodies results in chronic amnesic syndrome

       b.   Minimal gliosis in periaqueductal gray matter and gray matter of floor of fourth ventricle associated with residual ataxia

D.   Pathogenesis — due to acute **deficiency of thiamine**

   1.   Thiamine normally serves as cofactor for transketolase and two Krebs (tricarboxylic acid) cycle enzymes (pyruvate dehydrogenase complex and α-ketoglutarate dehydrogenase)

   2.   Immediate treatment with **high-dose parenteral thiamine** during *early* **Wernicke disease** produces rapid dramatic resolution of ocular palsies with slower improvement of ataxia and confusional state

   3.   Neuronal death follows **delay in thiamine administration** resulting in **Korsakoff's psychosis**

XI.   **Alcoholic cerebellar degeneration**

A.   **Truncal and gait ataxia** — slowly progressive truncal instability and incoordination of leg movements in chronic alcoholics

B.   Pathologic features

   1.   **Shrinkage of folia of superior (rostral) vermis and adjacent anterior lobe**

   2.   Loss of Purkinje cells, patchy loss of granule cells, and Bergmann gliosis (proliferation of Bergmann astrocytes normally found in Purkinje cell layer)

C. Presumably related to nutritional deficiency, but only some patients seem to improve following thiamine supplementation

XII. **Vitamin E (α-tocopherol) deficiency**

A. **Peripheral polyneuropathy** and **ataxia** (mimicking spinocerebellar degeneration), ophthalmoplegia, and pigmentary retinopathy

B. Pathologic features

1. Demyelination of spinal cord posterior columns

2. Dystrophic axons (terminal axonal enlargements containing accumulated filaments, membranes, abnormal mitochondria, and granular material)

3. Excessive lipofuscin pigmentation of neurons, astrocytes, and muscle cells

C. Associated with chronic **malabsorption syndromes** (such as occurs with cystic fibrosis, abetalipoproteinemia, liver disease, or intestinal resections) which result in deficiency of fat-soluble vitamins (vitamins A, D, E, K)

XIII. **Vitamin B$_{12}$ deficiency (pernicious anemia; subacute combined degeneration of spinal cord**; combined systems disease)

A. Clinical symptoms

1. Progressive moderate to severe **loss of posterior column sensation** (proprioception: vibratory and position sense); on examination, patient falls from standing position after eye closure (**positive Romberg test**)

2. **Spasticity** (due to spinal cord lateral column corticospinal tract involvement) with bilateral **extensor plantar responses** (Babinski reflexes) despite reduction or loss of tendon reflexes; can result in paraparesis or quadriparesis

3. Variable degrees of depression, memory disturbance, and **dementia** ("megaloblastic madness")

B. Diagnosis established by finding **elevated levels of methylmalonic acid in serum or urine, low serum vitamin B$_{12}$ levels,** or **positive Schilling test**; anemia (megaloblastic anemia) or disturbances of blood cell morphology (macrocytosis or hypersegmented neutrophils) not required for diagnosis

C. Pathologic features

    1. Vacuolation and fragmentation of myelin sheaths followed by axonal degeneration producing spongy appearance (spongiosis)

    2. Characteristically associated with conspicuous lack of astrocytic (glial) reaction; glial scarring can become evident in long-standing cases or after treatment with vitamin B_{12}

    3. **Involvement of spinal cord posterior and lateral columns, centered in mid-thoracic spinal cord**

    4. In severe cases, white matter of brain stem, cerebrum, and cerebellum is also involved

D. Most often due to autoimmune atrophic gastritis which results in **inadequate production of instrinsic factor** necessary for vitamin B_{12} absorption in distal ileum

E. Treatment after diagnosis consists of **parenteral vitamin B_{12} administration**; folic acid administration alone to a patient with vitamin B_{12} deficiency can worsen neurologic symptoms, while hematologic disturbances revert to normal

XIV. **Hypoglycemic encephalopathy**

A. Brain dysfunction related to critically **low blood glucose** levels (usually < 10 mg/dL)

B. Pathologic features — similar to those induced by hypoxia

C. Treatment requires **immediate intravenous glucose** administration (usually bolus of 50% glucose solution); early treatment can result in nearly complete recovery, but delayed treatment is associated with cerebral cortical neuronal destruction and residual neurologic signs

XV. **Hepatic encephalopathy**

A. Progressive **ataxia, dysarthria, asterixis (irregular flapping movements of outstretched arms)**, lethargy, stupor, and finally coma with characteristic EEG abnormalities

B. Results from **elevated blood ammonia levels** due to **chronic liver disease** and portosystemic shunting (from cirrhosis and portal hypertension)

    C.   Pathologic features

        1.   **Alzheimer type II astrocytosis** — astrocytes with enlarged vesicular (clear) nuclei; nuclei often appear paired

        2.   Alzheimer type II astrocytosis in gray matter (particularly deep layers of cerebral cortex, basal ganglia, and thalamus)

XVI. Uremic neuropathy

    A.   **Painful** distal sensorimotor polyneuropathy; often described as "burning feet"

    B.   Pathologic features — **distal axonal degeneration**

    C.   Does not correlate with degree of uremia; usually only responsive to restoration of normal renal function (renal transplantation)

XVII. Kernicterus

    A.   Disorder of neonates characterized by **bilirubin deposition** in brain due to passage of **unconjugated bilirubin** through **immature blood-brain barrier** compromised by hypoxia and acidosis

        1.   Full term neonates — associated with serum bilirubin levels greater than 20 mg/dL (usually from **erythroblastosis fetalis** due to Rh or ABO incompatibility)

        2.   Sick or premature newborns — relatively low serum bilirubin levels (10 mg/dL) can result in bilirubin deposition

    B.   Bilirubin deposition in basal ganglia, thalamus, hippocampus, inferior olivary nuclei, subthalamic nuclei, dentate nuclei, and most **cranial nerve nuclei** produces neuronal death and reactive glial scarring

    C.   Infants surviving neonatal period have **deafness** and **choreoathetoid cerebral palsy**

XVIII. Radiation effects

    A.   **Early-delayed effects**

        1.   Occur within several weeks of irradiation

166

2. Primary damage to oligodendrocytes with **demyelination** and necrosis of white matter, followed by some degree of remyelination

B. **Late-delayed effects**

1. Occur weeks, months, years after irradiation

2. **Hyalinization of blood vessel walls**

3. **Coagulative tissue necrosis** associated with **fibrinoid necrosis of vascular walls**

SUGGESTED ADDITIONAL READING

Victor M, Adams RD, Collins GH: *The Wernicke-Korsakoff Syndrome.* ed 2. Philadelphia, F. A. Davis Co, 1989.

Vinken PJ, Bruyn GW, Cohen MM, Klawans HL (eds): *Intoxications of the Nervous System-Part I. Handbook of Clinical Neurology. Volume 36.* Amsterdam, North-Holland Publishing Co, 1979.

Vinken PJ, Bruyn GW, Cohen MM, Klawans HL (eds): *Intoxications of the Nervous System-Part II. Handbook of Clinical Neurology. Volume 37.* Amsterdam, North-Holland Publishing Co, 1979.

Vinken PJ, Bruyn GW, Klawans HL (eds): *Metabolic and Deficiency Diseases of the Nervous System-Part I. Handbook of Clinical Neurology. Volume 27.* Amsterdam, North-Holland Publishing Co, 1976.

Vinken PJ, Bruyn GW, Klawans HL (eds): *Metabolic and Deficiency Diseases of the Nervous System-Part II. Handbook of Clinical Neurology. Volume 28.* Amsterdam, North-Holland Publishing Co, 1976.

Vinken PJ, Bruyn GW, Klawans HL (eds): *Metabolic and Deficiency Diseases of the Nervous System-Part III. Handbook of Clinical Neurology. Volume 29.* Amsterdam, North-Holland Publishing Co, 1977.

CHAPTER 11: PATHOLOGY OF NEUROMETABOLIC DISORDERS

I. General characteristics of genetic disorders

    A. Usually result from single defective **autosomal recessive** or **sex-linked recessive** gene

    B. Generally involve **complex metabolic pathways**

    C. **Storage disease** — deficiency of catabolic lysosomal enzyme (hence, **lysosomal storage disease**) resulting in accumulation of metabolites which cannot be further metabolized

II. **Gaucher's disease**

    A. Autosomal recessive inheritance due to defective gene localized to long arm of chromosome 1 (region 1q21)

    B. Accumulation of glucocerebroside due to deficiency of enzyme **glucocerebroside-ß-glucosidase** which normally cleaves glucose from ceramide backbone

    C. Clinical features

        1. **Type I (adult form**; chronic non-neuronopathic form)

            a. Most common form of Gaucher's disease, usually affecting **Ashkenazic** Jews

            b. Clinical presentation

                (1) **Hepatosplenomegaly** with signs of **hypersplenism** (anemia, leukopenia, thrombocytopenia) and **hepatic dysfunction** (prolonged prothrombin time and abnormal liver function tests)

        (2)   **Skeletal deformities** and episodes of severe bone pain

        (3)   No neurologic signs

2.   **Type II (infantile form**; acute neuronopathic form)

    a.   **No ethnic group predilection**

    b.   Clinical presentation

        (1)   **Hepatosplenomegaly, spasticity, developmental delay,** strabismus, and upper motor neuron brain stem signs (including supranuclear gaze palsy and pseudobulbar palsy) developing during first several months of life

        (2)   Usually fatal by age 18 months

3.   **Type III (juvenile form**; subacute neuronopathic form)

    a.   Symptoms intermediate between type I and type II forms

    b.   Two types of clinical presentation:

        (1)   **Supranuclear gaze palsy** and **hepatosplenomegaly** along with varying degrees of progressive cognitive decline (**dementia**) and seizures beginning in late adolescence

        (2)   Supranuclear gaze palsy with severe progressive hepatosplenomegaly and death from liver failure

D. Pathologic features

    1.   Markedly **elevated serum acid phosphatase levels**

    2.   **Gaucher cells**

        a.   Large **foamy vacuolated macrophages** (reticuloendothelial cells) particularly prominent in spleen, liver, bone marrow, and lymph nodes

        b.   **Enlarged lysosomes** contain stored material which has electron microscopic appearance of parallel layers of membranes

3. Numerous Gaucher cells in perivascular (Virchow-Robin) spaces in brain in all forms

4. Prominent neuronophagia in type II form

E. Therapy — intravenous infusion of modified human placental gluco-cerebrosidase is taken up by macrophages reversing signs and symptoms in type I disease, but does not affect neurologic abnormalities in type II or type III disease

III. Niemann-Pick disease

A. Two distinct groups of autosomal recessive disorders characterized by **accumulation of sphingomyelin**

1. **Group I (type A and type B Niemann-Pick disease)** — due to defective gene on short arm of chromosome 11 (region 11p15) resulting in **deficiency of enzyme sphingomyelinase** which normally cleaves ethanolamine from ceramide backbone

2. **Group II (type C and type D Niemann-Pick disease) — aberrant intracellular cholesterol homeostasis**

B. Clinical features

1. Type A (**acute neuronopathic** form)

    a. Common among Ashkenazic Jews

    b. **Hepatosplenomegaly** developing in first months of life, brownish-yellow skin discoloration, generalized lymphadenopathy, vomiting and feeding difficulties, **blindness** (often with macular cherry-red spot), spasticity, and progressive neurologic deterioration

    c. Usually fatal before age 2 years

2. Type B (**visceral form**)

    a. **Hepatosplenomegaly** with signs of liver dysfunction and diffuse **pulmonary infiltrates** with recurring respiratory infections

b. **Hyperlipidemia** with increased serum triglyceride and low-density lipoprotein (LDL) levels and decreased high-density lipoprotein (HDL) levels

c. **No neurologic abnormalities**

3. **Type C and Type D**

    a. Slight to moderate **hepatosplenomegaly** (prolonged neonatal jaundice common)

    b. Variable age of onset (newborn to adult) of progressive neurologic signs of **vertical supranuclear gaze palsy, extrapyramidal movement disorders, ataxia,** spasticity, seizures, visual impairment, and dysphagia (leading to recurrent aspiration pneumonia)

    c. Usually fatal after 10 to 20 years

    d. Type D patients all have common Nova Scotian (Acadian) ancestry

C. Pathologic features

1. **Type A**

    a. **Niemann-Pick cells (foam cells)**

        (1) Large (up to 100 μm diameter) foamy autofluorescent lipid-laden (often multinucleate) macrophages in liver, spleen, lymph nodes, and bone marrow

        (2) **Enlarged lysosomes** contain lamellar myelin-like figures

    b. **Brain atrophy with marked neuronal loss and gliosis;** remnant neurons have foamy cytoplasm due to stored material similar to that in macrophages; foamy macrophages present throughout leptomeninges and perivascular (Virchow-Robin) spores

2. **Type B** — visceral changes similar to type A, but no nervous system involvement

3. **Type C and Type D**

    a. Foamy vacuolated autofluorescent lipid-laden macrophages in spleen, liver, and bone marrow ("sea-blue histiocytes")

    b. Brain atrophy with generalized neuronal loss and gliosis; remnant neurons are "ballooned" with foamy cytoplasm containing enlarged lysosomes with lamellar inclusions

IV. G_{M2}-**gangliosidosis**

A. **Prototype autosomal recessive lipidosis**

    1. **Deficiency of lysosomal hexosaminidase activity**

        a. Hexosaminidase A — dimer of one α-subunit and one ß-subunit

        b. Hexosaminidase B — dimer of two ß-subunits

        c. Gene for α-subunit localized to long arm of chromosome 15 (region 15q23-24) and gene for ß-subunit localized to long arm of chromosome 5 (region 5q11-13)

    2. **Hexosaminidase A deficiency** common in **Ashkenazic Jews** with carrier rate of about 1 in 30; **hexosaminidase A and B deficiency** common in **Arab population**

B. Clinical features

    1. **Deficiency of hexosaminidase A**

        a. **Tay-Sachs disease** — initially normal infant; by age 3 months, infant develops **hyperacusis** (exaggerated startle to sound), hyperexcitability, psychomotor retardation, hypotonia, **macular cherry-red spot**; later progressive decerebrate posturing and **opisthotonus** (spasmodic posturing with head and spine bent backwards), **macrocrania** (**head enlargement**), and seizures; usually fatal before age 2 years

        b. **Late-onset G_{M2}-gangliosidosis** — initially clumsy child with **progressive proximal weakness** beginning in adolescence; ataxia and behavior disturbances developing in early adulthood; electrophysiologic evidence of **peripheral neuropathy** with widespread denervation

2. **Deficiency of hexosaminidase A and B (Sandhoff disease)** — similar clinical presentations as Tay-Sachs disease and late-onset G_{M2}-gangliosidosis

C. **Pathologic features**

  1. **Tay-Sachs disease**

  a. Initial brain enlargement (50% increase in brain weight) followed later by marked cerebral cortical atrophy and loss of white matter

  b. **Ballooned neurons** throughout brain and spinal cord, peripheral ganglia, myenteric plexus, and adrenal medulla containing enlarged lysosomes with granular and lamellar storage material

  c. **Meganeurites — torpedo-shaped swellings** interspersed between neuronal cell body and axon; aberrant synaptic connections with meganeurites responsible for hyperacusis, hyperexcitability, and seizures

  2. **Sandhoff disease**

  a. Neuropathologic changes similar to Tay-Sachs disease

  b. Foamy macrophages and vacuolated parenchymal cells in many visceral organs including kidneys, pancreas, lungs, spleen, liver, lymph nodes, and bone marrow

V. **Fabry's disease**

A. **X-linked recessive disorder** due to genetic defect localized to long arm of X-chromosome (Xq21-22 region)

  1. **Only sphingolipidosis with sex-linked inheritance pattern**

  2. **Deficiency of α-galactosidase A** (trihexosylceramide α-galactosidase)

  3. Enzyme deficiency detectable in plasma, leukocytes, tears, and cultured cells

B. Clinical presentation

  1. Onset in **childhood or early adolescence** of **paroxysmal excruciating burning pain** in hands and feet (acroparesthesias),

sensory loss, anhydrosis, unexplained fevers, characteristic corneal and lens opacities, and **progressive renal failure**

2. **Angiokeratoma corporis diffusum — cutaneous telangiectases** which are hyperkeratotic dark-red to blue lesions in bathing trunk distribution

C. Pathologic features

1. Accumulation of **ceramide trihexoside** and other glycolipids with terminal α-galactose; blood group B substance also accumulates in patients with blood group B or AB resulting in more severe disease

2. Accumulation in lysosomes of vascular endothelial, perithelial, and smooth muscle cells throughout body results in progressive vascular stenosis and occlusion leading to ischemia and infarction

3. **Ballooning of neurons** (containing enlarged lysosomes) in spinal dorsal root and autonomic ganglia, hypothalamus, intermediolateral cell column, and brain stem reticular formation; peripheral nerves show preferential loss of small fibers

4. Lysosomes contain lamellar inclusions by electron microscopy

D. Treatment — intravenous infusion of purified enzyme or renal transplantation with normal kidney (that can produce normal enzyme) produces biochemical and some clinical improvement

VI. **Ceroid lipofuscinoses**

A. Autosomal recessive inheritance

B. Metabolic abnormalities not completely defined and deficient enzyme not yet identified; abnormality of metabolism of long chain polyunsaturated fatty acids in infantile form

C. Clinical types

1. Infantile (**Hagberg-Santavuori**) — rapidly progressive retardation and seizures beginning late in first year; microcephaly and death before age 10 years

2. Late infantile (**Bielschowsky-Jansky**) — relatively normal until onset of seizures between ages 4 to 9 years; slow progression with loss of vision; death before age 10 years

3. Early juvenile (**Lake-Cavanaugh**) and juvenile (**Batten-Mayou, Spielmeyer-Vogt**) — loss of vision with pigmentary retinopathy between ages 4 to 9 years; slow progression with death in late teens or early twenties

4. Adult (**Kufs**) — progressive dementia, behavioral disturbances from childhood, seizures, dysarthria, extrapyramidal and cerebellar signs; no visual symptoms

D. Pathologic features

1. Autofluorescent storage material (**ceroid lipofuscin**) in macrophages in spleen, liver, lymph nodes, and lamina propia of gastrointestinal tract and in smooth muscle cells of gastrointestinal tract, arterioles and arteries, skeletal and cardiac muscle, renal glomeruli and tubules, and other organs.

2. Neurons typically not ballooned despite containing increased amounts of autofluorescent ceroid lipofuscin; brain atrophy common at all ages, most severe in younger ages; dura thickened

3. Electron microscopy shows inclusions typical for each type

VII. **Metachromatic leukodystrophy**

A. Autosomal recessive inheritance due to defective gene localized to long arm of chromosome 22 (region 22q13)

B. Accumulation of **sulfatide** (galactocerebroside sulfate) due to deficiency of enzyme **arylsulfatase A** which normally cleaves sulfate from sulfatide

C. Clinical features

1. Late infantile — progressive gait disturbance in toddler (age 2 years) followed by truncal instability, visual loss (and optic atrophy), peripheral neuropathy with flaccid weakness, quadriplegia; fatal after 4-5 years

2. Juvenile — progressive gait disturbance and diminished school performance beginning around age 6 years followed by spasticity,

peripheral neuropathy, visual loss, quadriplegia, and death by late adolescence

    3.   Adult — insidious onset in late adolescence or early adulthood of slowly progressive gait disturbance, behavioral problems, spasticity, and peripheral neuropathy

  D. Pathologic features

    1.   Firm gray-brown or chalky, focally-cavitated white matter (with relative sparing of subcortical U-fibers) in brain and spinal cord

    2.   Marked loss of oligodendrocytes and myelin with granular masses of accumulated sulfatide in remnant oligodendrocytes and macrophages

    3.   Segmental demyelinization in peripheral nerves with accumulated sulfatide in remnant Schwann cells and endoneurial macrophages

        a.   Sulfatide accumulations stain strongly with periodic acid-Schiff (PAS) and stain metachromatically with toluidine blue or cresyl violet

        b.   **Metachromasia** — color change when dye binds to certain compounds (distinguished from orthochromasia in which dye retains original color)

    4.   Enlarged lysosomes contain lamellar inclusions

    5.   Sulfatide accumulation in renal tubular cells, hepatic Kupffer cells, adrenal medulla, islets of Langerhans, and gall bladder epithelium

VIII. **Krabbe's disease (globoid cell leukodystrophy)**

  A. Autosomal recessive inheritance; defective gene localized to chromosome 14

  B. Accumulation of **galactocerebroside** (and pyschosine) due to deficiency of enzyme **galactocerebroside-ß-galactosidase** which normally cleaves galactose from ceramide backbone and from sphingosine

  C. Clinical features

    1.   Onset before age 6 months of loss of developmental milestones, irritability, spasticity, and recurrent unexplained fever followed by seizures, opisthotonus, blindness and death usually before age 2 years

2. Elevated cerebrospinal fluid protein levels, microcephaly, and slow nerve conduction velocities

D. Pathologic features

 1. Small firm brain with discolored and cavitated white matter (except preserved U-fibers) of cerebrum, brain stem, cerebellum, and spinal cord

 2. Neurons generally not affected

 3. Wide spread loss of myelin; marked gliosis with presence of numerous perivascular clusters of **globoid cells** (large multinucleated macrophages filled with accumulated PAS-positive galactocerebroside)

 4. Segmental demyelination in peripheral nerves; galactocerebroside accumulation in Schwann cells

 5. Enlarged lysosomes contain elongated crystalline or tubular structures

 6. **Psychosine** (galactosylsphingosine) accumulation — even small amounts are highly toxic for oligodendrocytes resulting in cell death with consequent myelin destruction and impaired myelin formation

IX. **Peroxisomal disorders**

A. **Adrenoleukodystrophy** (Schilder's disease; Addison's disease with leukodystrophy)

 1. X-linked recessive disorder due to genetic defect localized to long arm of X-chromosome (Xq28)

 2. Abnormal metabolism of very long-chain (24 to 30 carbon chain length) fatty acids which are normally metabolized by peroxisomal fatty-acid ß-oxidation; **single enzyme defect** with normal number of peroxisomes

 3. Clinical features

  a. Childhood cerebral form

   (1) **Behavioral disturbances** beginning around age 4 years with visual and hearing disturbances followed by progressive **spasticity**, dysphagia, **blindness**, and death within several years

(2) **Adrenal insufficiency — skin hyperpigmentation** and impaired cortisol response to adrenocorticotrophic hormone (ACTH)

b. Adult cerebral form — behavioral disturbances beginning in young adulthood progressing to dementia, seizures, spasticity, dysphagia, blindness and death within several years

c. Adrenomyeloneuropathy

(1) Progressive spasticity (**spastic paraparesis**), lower extremity weakness and sensory loss, and impaired bladder function

(2) **Adrenal insufficiency** and low serum testosterone levels

(3) Subtle behavioral symptoms

d. Symptomatic heterozygotes — some heterozygous females have late-onset slowly progressive spastic paraparesis with fluctuating symptoms

4. Pathologic features

a. Symmetrical **demyelination of cerebral white matter** which is **most severe in occipital regions** with sparing of U-fibers; gliosis; prominent perivascular inflammation; neurons spared

b. **Adrenomyeloneuropathy** — loss of myelinated fibers and oligodendrocytes in spinal cord due to distal axonopathy with greatest losses in lumbar corticospinal and cervical gracile and dorsal spinocerebellar tracts

5. Originally termed Schilder's disease, which may actually represent childhood multiple sclerosis

B. **Zellweger's syndrome (cerebrohepatorenal syndrome)**

1. Autosomal recessive inheritance

2. **Defective peroxisomal biogenesis** with resultant **multiple enzyme deficiencies** producing abnormalities in multiple systems including fatty acid ß-oxidation, bile acid synthesis, cholesterol biosynthesis, plasmalogen synthesis, and amino acid metabolism

3. Clinical features — **floppy dysmorphic infant** with high narrow forehead, round cheeks, flat root of nose, wide-set eyes, puffy eyelids, corneal opacities, pursed lips, narrow high-arched palate and small chin; poor suck and swallowing; congenital heart disease (septal defects or patent ductus arteriosus); jaundice; splenomegaly; cystic dysplasia of kidneys; genital anomalies

4. Pathologic features

    a. Abnormal neuronal migration with consequent abnormally small gyri (microgyria), abnormally large gyri (pachygyria), glial and neuronal heterotopias, abnormally placed (heterotopic) Purkinje cells, and dysplastic inferior olivary nuclei

    b. **Absence of peroxisomes in liver and kidney**; cystic dysplasia of kidneys, cardiac ventriculoseptal defects, and cirrhosis

    c. Peroxisome ghosts — empty membranous sacs identifiable ultrastructurally

X. **Mucopolysaccharidoses**

A. Disorders resulting in lysosomal **accumulation of glycosaminoglycans**

B. Autosomal recessive inheritance except for X-linked (sex-linked) inheritance in Hunter syndrome

C. Clinical features — characterized by **skeletal anomalies with nervous system abnormalities in some forms**

    1. **Hurler, Scheie, Hurler-Scheie — dwarfism, grotesque facial features**, protuberant abdomen, joint contractures, mental retardation

    2. **Hunter** — skeletal deformities, mental retardation

    3. **Sanfilippo** — severe mental retardation, minor skeletal abnormalities

    4. **Morquio** — skeletal deformities, no mental retardation

    5. **Maroteaux-Lamy** — skeletal deformities, no mental retardation, corneal opacity

D. Urinary excretion of dermatan sulfate, heparan sulfate, or keratan sulfate detectable

E. Pathologic features

1. Each clinical type has characteristic constellation of organ involvement (i.e., deafness and cardiac abnormalities are found to varying degrees in both Hurler and Hunter types)

2. Stored mucopolysaccharides are found in most organs and tissues

3. Shared biochemical pathways for gangliosides and mucopolysaccharides results in storage of gangliosides in neurons and liver macrophages

4. Alteration in brain **correlates with degree of mental retardation** (i.e., severe in Hurler and Sanfilippo syndromes and less severe in Hunter syndrome)

   a. Thickened leptomeninges, meningeal cysts, increased perivascular spaces in white matter, and stored material in neurons

   b. Increased mucopolysaccharides are present in meninges and perivascular spaces with gangliosides present in neurons

DISEASE	ENZYME	ACCUMULATED MATERIAL
Hurler	α-L-Iduronidase	Dermatan sulfate, heparan sulfate
Hurler-Scheie	α-L-Iduronidase	Dermatan sulfate
Scheie	α-L-Iduronidase	Dermatan sulfate
Hunter (A&B)	Iduronosulfate sulfatase	Dermatan sulfate, heparan sulfate
Sanfilippo A	Sulfamidase	Heparan sulfate
Sanfilippo B	α-N-Acetylglucosaminidase	Heparan sulfate
Sanfilippo C	Acetyl-CoA:α-glucosaminide N-acetyltransferase	Heparan sulfate
Sanfilippo D	N-Acetylglucosamine-6-sulfate sulfatase	Heparan sulfate
Morquio A	N-Acetylgalactosamine-6-sulfate sulfatase	Keratan sulfate, chondroitin-6-sulfate
Morquio B	ß-Galactosidase	Keratan sulfate
Maroteaux-Lamy	Acetylgalactosamine-4-sulfate sulfatase	Dermatan sulfate
ß-Glucuronidase Deficiency	ß-Glucuronidase	Chondroitin-4/6-sulfate, dermatan sulfate, heparan sulfate

XI.   **Aminoacid disorders**

A.   **Phenylketonuria**

1.   Autosomal recessive inheritance due to defective gene on long arm of chromosome 12

2.   Absence of enzyme phenylalanine hydroxylase which normally converts phenylalanine to tyrosine resulting in **accumulation of phenylalanine**

3.   Clinical features — infants present with seizures, restlessness, muscular hypertonicity, microcephaly, abnormal body movements, severe mental retardation; defective pigment formation results in fair skin, blond hair, and blue eyes

4.   **Excess phenylalanine converted to phenylpyruvate** (phenylketone) and phenylacetate (odor producing metabolite) which can be detected in urine using ferric chloride test

5.   Pathologic features — microcephaly; in patients less than age 5 years, **gliosis and spongiosis of white matter, and delay in myelination**; over age 5 years, less specific foci of spongiosis, myelin pallor, gliosis, and myelin breakdown; impaired dendritogenesis of neurons

6.   Treatment — restriction of dietary phenylalanine intake

B.   **Homocystinuria**

1.   Autosomal recessive inheritance due to defective gene localized on long arm of chromosome 21 (region 21q22) resulting in absence of **liver enzyme cystathionine ß-synthase** which combines homocysteine and serine to form cystathionine (which is then broken down into cysteine and homoserine)

2.   Clinical features

a.   Connective tissue abnormalities include **ectopia lentis** and **osteoporosis**

b.   **Thrombotic features** include coronary and carotid occlusion, renal artery stenosis with hypertension and thromboses of large veins such as the inferior vena cava

c.   Mental retardation, seizures, and cerebrovascular thrombosis

3. Detected by abnormal urinary cyanide nitroprusside test; elevation of urinary homocystine and plasma methionine and homocystine

4. Pathologic features due to ischemic brain injury

5. Treatment — large doses (up to 1000 mg/day) of pyridoxine (vitamin B_6) which is cofactor for cystathionine ß-synthase useful in those patients with residual enzyme activity; restriction of dietary methionine intake useful in pyridoxine nonresponders

XII. **Transferase-deficient galactosemia**

A. Autosomal recessive disorder characterized by absence of enzyme **galactose-1-phosphate uridyl transferase** which is part of three-step conversion of galactose to glucose

B. Clinical features — infants fed milk develop hepatosplenomegaly, jaundice, hypoglycemia, cataracts, failure to thrive, mental retardation

C. Abnormal liver function tests, elevated blood galactose, galactosuria, and acidosis

D. Pathologic features — rare autopsy studies have shown neuronal loss and reactive gliosis

E. Treatment — restriction of dietary galactose intake (for example, feeding of soy-based milk substitutes)

XIII. **Glycogen storage disorders**

A. **Acid maltase deficiency**

1. Autosomal recessive inheritance due to defective gene on long arm of chromosome 17 (region 17q23) resulting in deficiency of lysosomal enzyme acid **α-glucosidase**

2. Clinical types

a. **Pompe's disease** (infantile acid maltase deficiency; generalized glycogenesis)

(1) Neonatal presentation of **diffuse hypotonia and weakness ("floppy infant")** despite normal or increased

muscle bulk, **macroglossia, massive cardiomegaly, hepatosplenomegaly**; usually fatal by age 1 year

    (2)   Pathologic features

        (a)   **Vacuolar myopathy** due to massive cytoplasmic accumulation of glycogen which stains positively with periodic acid-Schiff (PAS)

        (b)   Glycogen accumulation in spinal cord anterior horn cells and brain stem motor neurons explains flaccid quadriplegia

        (c)   Glycogen is present in enlarged lysosomes and free in large intracellular pools

  b.   Myopathy of childhood or adulthood

    (1)   **Slowly progressive muscle weakness** with particular **involvement of respiratory muscles** resulting in death from ventilatory insufficiency

    (2)   **Calf enlargement** can mimic muscle pseudohypertrophy of Duchenne muscular dystrophy

    (3)   **Electromyography** shows denervation (fibrillations) and **myotonic discharges**

    (4)   Pathologic features — **vacuolar myopathy**

B. **Muscle phosphorylase deficiency** (McArdle's disease)

  1.   Autosomal recessive inheritance due to defective gene on long arm of chromosome 11 (region 11q13)

  2.   **Absence of muscle isoenzyme form of phosphorylase**

    a.   Phosphorylase normally initiates glycogen breakdown by removing α-1,4-glucosyl residues from outer branches of glycogen

    b.   Phosphorylase-limit-dextrin (PLD) — partially digested glycogen molecule after phosphorylase action; peripheral chains have only four remaining glucosyl residues which are acted upon by debrancher enzyme (which removes those remaining four glucosyl residues)

3. Clinical features

    a. Exercise intolerance with premature fatigue and muscle pain

    b. Exercise-induced painful muscle contractures (stiffening; "electrically-silent cramps")

    c. Severe **post-exercise muscle necrosis (rhabdomyolysis), elevated serum creatine kinase** (CK), and **myoglobinuria** (which can produce renal failure)

    d. Reduction in symptoms with high carbohydrate loading prior to and during exercise

    e. **"Second wind" phenomenon** — increased exercise tolerance following brief pause and then resumption of activity; results from increased delivery of blood-borne free fatty acids and glucose to exercising muscle

4. Pathologic features

    a. **Glycogen accumulation** in muscle fibers

    b. Necrotic muscle fibers identifiable in exercised muscle

C. **Muscle phosphofructokinase deficiency**

1. Autosomal recessive inheritance due to defective gene on chromosome number 1

2. **Absence of muscle isoenzyme subunit of phosphofructokinase** which is rate-limiting enzymatic step in glycolysis (metabolism of glucose to pyruvate)

3. Clinical symptoms similar to muscle phosphorylase deficiency

    a. Fasting improves exercise tolerance due to increase in blood levels of free fatty acids

      b.   **"Out of wind" phenomenon** — worsening of symptoms following carbohydrate loading (which reduces blood levels of free fatty acids)

XIV. Wilson's disease (hepatolenticular degeneration)

A.   Autosomal recessive inheritance of genetic defect localized to long arm of chromosome 13 (region 13q14-21)

B.   Pathophysiology

    1.   Deficient enzyme not identified

    2.   **Low serum ceruloplasmin and copper levels** (in some but not all patients)

    3.   **Defective copper metabolism**

    4.   Persistent aminoaciduria

    5.   **Elevated tissue copper levels**

C.   Clinical features

    1.   Progressive ataxia, choreoathetosis, dystonia, dysarthria, and hepatic dysfunction (cirrhosis)

    2.   **Kayser-Fleischer ring** on slit-lamp examination

D.   Pathologic features

    1.   Hepatic **cirrhosis**

    2.   **Striatum (putamen and caudate) shrunken** with brown discoloration; cavitation of putamen; spongy softening of deep white matter

    3.   Marked loss of neurons in striatum

    4.   **Alzheimer type II astrocytosis** in putamen, caudate, subthalamic nucleus, thalamus

    5.   Opalski cells — large, rounded astrocytes with granular cytoplasm found in thalamus, globus pallidus, substantia nigra

E. Treatment — low copper diet and chelation therapy with D-penicillamine to remove accumulated excess tissue copper and prevent reaccumulation

XV. **Alexander's disease**

A. Sporadic disorder with no known enzymatic defect

B. Clinical features — infant with head enlargement, progressive loss of developmental milestones, and seizures

C. Pathologic features — diffuse demyelination, along with innumerable **Rosenthal fibers** around blood vessels and beneath pia

XVI. **Canavan's disease (spongy degeneration; van Bogaert-Bertrand disease)**

A. Autosomal recessive inheritance

B. **Absence of enzyme aspartoacylase** which normally is localized to brain white matter and hydrolyzes N-acetyl-L-aspartic acid to L-aspartic acid and acetate

C. Clinical features — developmental delay evident after age 3 months, hypotonia, macrocephaly, seizures, optic atrophy, and spasticity; usually fatal by age 10 years

D. Pathologic features

1. Accumulation of N-acetyl-L-aspartic acid in brain (particularly white matter)

2. Gelatinous rarefied white matter containing innumerable vacuoles (**"spongy degeneration"**)

3. Proliferation of astrocytes (Alzheimer type II astrocytes)

XVII. **Leigh's disease** (infantile subacute necrotizing encephalopathy)

A. Uncertain inheritance of abnormality in mitochondrial function with resultant deficiency of complex IV (cytochrome c oxidase), often along with deficiency of pyruvate dehydrogenase

B. Clinical features

    1. Normal development up to age 6 to 12 months, with subsequent **arrest of psychomotor development**, feeding problems, swallowing difficulty, **ataxia, optic atrophy** (loss of vision), ophthalmoplegia, nystagmus, **hypotonia**, and **breathing difficulty** (including apnea)

    2. **Rarefaction (low density) of basal ganglia** (especially putamen) on computed tomographic (CT) scan or magnetic resonance imaging (MRI)

    3. **Elevated serum lactate** and pyruvate levels

C. Pathologic features

    1. Ragged red fibers and **abnormal mitochondria** in muscle biopsy

    2. Focal symmetrical necrosis with microcystic vacuolation and capillary proliferation in basal ganglia, thalamus, brain stem, and spinal cord (especially dorsal columns)

XVIII. **Lesch-Nyhan disease**

A. Sex-linked recessive inheritance due to abnormal gene localized to long arm of X-chromosome (Xq2)

B. **Defective purine metabolism** due to **absence of enzyme hypoxanthine-guanine phosphoribosyltransferase** which is primary **salvage enzyme** catalyzing conversion of hypoxanthine to inosine monophosphate (IMP) or guanine to guanine monophosphate (GMP) by reaction with 5-phosphoribosyl-1-pyrophosphate

C. Clinical features

    1. Severe **mental retardation**, spasticity, choreoathetosis, **self-mutilating behavior**

    2. Hyperuricemia, crystalluria, nephrolithiasis, and death from renal failure in second or third decade of life

D. Pathologic features

    1. Gout, tophaceous deposits, nephrolithiasis and associated renal alterations

187

2. No gross or microscopic nervous system changes, but decreased levels of brain catecholamines (particularly dopamine)

XIX. **Porphyria**

A. Group of disorders with **defective heme synthesis** characterized by acute attacks

B. **Autosomal dominant pattern of inheritance**; affected individuals have 50% of normal enzyme activity

C. Enzymatic defects

1. **Acute intermittent porphyria** — porphobilinogen deaminase

2. **Hereditary coproporphyria** — coproporphyrinogen oxidase

3. **Variegate porphyria** — protoporphyrinogen oxidase

D. Clinical features

1. **Episodes of acute abdominal pain**, constipation, and vomiting suggestive of peritonitis but without fever, leukocytosis or rebound tenderness

2. Psychosis, delirium, seizures, motor neuropathy (with proximal weakness)

3. Episodes precipitated by drug administration or metabolic changes:

a. Heme biosynthesis is normally regulated by end product inhibition (concentration of heme either stimulating or inhibiting enzymes involved with its synthesis);

b. Up-regulation of heme biosynthesis follows increased heme utilization — for example, as occurs to meet increased cellular demand for mitochondrial cytochromes in association with:

(1) Various drugs — including **barbiturates** and other anticonvulsants, **sedative-hypnotics**, **sulfonamides**, **alcohol**, and chloramphenicol

(2) Fluctuations in levels of sex steroid hormones (estrogen, progesterone, testosterone)

188

(3) **Starvation** or **low carbohydrate intake** (as in dieting or with illness)

c. Partial blockade of heme biosynthesis results in energy crises in cells (due to lack of sufficient heme proteins) and marked increase in synthesis of heme precursors (particularly δ-aminolevulinic acid and porphobilinogen)

d. **Increased urinary excretion of porphobilinogen** detectable in **Watson-Schwartz test**

4. Skin rash in variegate porphyria and hereditary coproporphyria

E. Pathologic features

1. Reduced number of fibers in distal nerves, axonal atrophy, distal axonal degeneration, motor fibers consistently more affected than sensory

2. Degeneration of dorsal columns and central chromatolysis of motor neurons

SUGGESTED ADDITIONAL READING

Beutler E: Gaucher's disease. *New Engl J Med* 1991;7:1354-1360.

Dyken PR: The neuronal ceroid lipofuscinoses. *J Child Neurol* 1989;4:165-174.

Goebel HH, Braak H: Adult neuronal ceroid-lipofuscinoses. *Clin Neuropathol* 1989;8:109-119.

Naidu S, Moser HW: Peroxisomal disorders. *Neurol Clin* 1993;8:507-528.

Powers JM, Tummons RC, Caviness, Jr VS, Moser AB, Moser HW: Structural and chemical alterations in the cerebral maldevelopment of fetal cerebro-hepato-renal (Zellweger) syndrome. *J Neuropath Exp Neurol* 1989;48:270-289.

Rosenberg RN, Prusiner SB, DiMauro S, Barchi RL, Kunkel LM (eds): *The Molecular and Genetic Basis of Neurological Disease*. Boston, Butterworth-Heinemann, 1993.

Volk BW, Adachi M, Schneck L: The gangliosidoses. *Human Pathol* 1975;6:555-569.

Summary of Selected Neurometabolic Disorders

DISEASE	ENZYME DEFICIENCY	ACCUMULATING SUBSTANCE	CLINICAL SYMPTOMS
Alexander's disease	[deficient enzyme not defined]	Rosenthal fibers	Progressive psychomotor retardation and head enlargement; diffuse demyelination with numerous perivascular and subpial Rosenthal fibers
Adrenoleukodystrophy	[peroxisomal enzyme not yet defined]	Very long chain fatty acids (C_{24-30})	Behavior disturbance, visual loss, and spasticity; generalized demyelination beginning in occipital lobes; adrenal insufficiency
Canavan's disease	Aspartoacylase	N-acetyl-L-aspartic acid	Developmental delay, hypotonia, macrocephaly, seizures, optic atrophy, spasticity; spongy degeneration of white matter, Alzheimer type II astrocytosis
Ceroid lipofuscinosis	[deficient enzyme not defined]	Ceroid lipofuscin	Mental retardation (dementia in adult form), seizures, pigmentary retinopathy
Fabry's disease	Ceramide trihexoside α-galactosidase	Ceramide trihexoside	Renal failure, rash in bathing trunk distribution; painful peripheral neuropathy; X-linked inheritance
Galactosemia	Galactose-1-phosphate uridyl transferase	Galactose	Hepatosplenomegaly, jaundice, hypoglycemia, cataracts, failure to thrive, mental retardation in milk-fed infants
Gaucher's disease	Glucocerebroside β-glucosidase	Glucocerebroside	Hepatosplenomegaly, bony erosion; mental retardation in infantile form; foamy vacuolated (Gaucher) cells in bone marrow, liver, spleen
Homocystinuria	Liver cystathionine β-synthase	Homocystine; methionine	Long limbs, lens dislocation, renal artery stenosis and hypertension, mental retardation, seizures; propensity to vascular thromboses with consequent cerebral ischemia (stroke)
Krabbe's disease (globoid cell leukodystrophy)	Galactocerebroside β-galactosidase	Galactocerebroside	Mental retardation; absent myelin; white matter globoid cells
Leigh's disease	Cytochrome c oxidase; pyruvate dehydrogenase	——————	Psychomotor arrest, hypotonia, ataxia, optic atrophy, ophthalmoplegia; hyperlactatemia and hyperpyruvatemia; cavitation of basal ganglia
Lesch-Nyhan syndrome	Hypoxanthine-guanine phosphoribosyltransferase	Uric acid	Severe mental retardation, spasticity, choreoathetosis, self-mutilation; gout, tophaceous deposits, urate kidney stones, renal damage; decreased levels of brain catecholamines (particularly dopamine)
Metachromatic leukodystrophy	Arylsulfatase A	Sulfatide	Gait disturbance, peripheral neuropathy; diffuse demyelination; accumulation of metachromatic material
McArdle's disease	Muscle phosphorylase	Glycogen	Exercise intolerance, exercise-induced muscle contractures, post-exercise muscle necrosis (rhabdomyolysis), myoglobinuria
Niemann-Pick disease (group I)	Sphingomyelinase	Sphingomyelin	Hepatosplenomegaly, mental retardation; foam cells in viscera; ballooned neurons
Phenylketonuria	Phenylalanine hydroxylase	Phenylpyruvic acid	Mental retardation, seizures, microcephaly; delayed myelination, gliosis, impaired dendritogenesis
Phosphofructokinase deficiency	Phosphofructokinase	Glycogen; glycolytic intermediates	Exercise intolerance, exercise-induced muscle contractures, post-exercise muscle necrosis (rhabdomyolysis), myoglobinuria
Pompe's disease (acid maltase deficiency)	Acid α-glucosidase	Glycogen	Floppy infant with macroglossia, massive cardiomyopathy, hepatosplenomegaly; vacuolar myopathy
Porphyria (acute intermittent form)	Porphobilinogen deaminase	δ-Aminolevulinic acid; porphobilinogen	Episodic acute abdominal pain, psychosis, delirium, seizures, motor neuropathy
Sandhoff disease	Hexosaminidase A and B	G_{M2}-ganglioside and globoside	Similar to Tay-Sachs disease; ballooned neurons containing stored material
Tay-Sachs disease	Hexosaminidase A	G_{M2}-ganglioside	Mental retardation; macular cherry-red spot; seizures; early head enlargement; ballooned neurons containing storage material
Wilson's disease (hepatolenticular degeneration)	Defective copper metabolism [deficient enzyme not defined]	Copper	Progressive ataxia, choreoathetosis, dystonia, dysarthria, hepatic dysfunction (cirrhosis); cavitation of putamen, Alzheimer type II astrocytosis of basal ganglia and thalamus
Zellweger's (cerebro-hepatorenal) syndrome	Multiple enzymes (absence of peroxisomes)	Very long chain fatty acids (C_{24-30})	Floppy dysmorphic infant; corneal opacities; jaundice; cystic dysplasia of kidneys; neuronal migration defects with heterotopias

CHAPTER 12: PATHOLOGY OF DEMYELINATING DISORDERS

I.   Differentiation of demyelinating disease, dysmyelinating disease, and secondary demyelination

    A.   **Demyelinating disease** (myelinoclastic disease)

        1.   **Myelin sheath destruction** — acquired disease process of central or peripheral nervous system in which normally formed myelin sheaths are destroyed while axons are spared

        2.   **Sudanophilia** — macrophage accumulation of **neutral fats** (triglycerides and cholesterol esters) resulting from breakdown of myelin sheath imparts sudanophilic staining properties (positive staining with Sudan black B stain)

        3.   Disease processes — multiple sclerosis, acute disseminated encephalomyelitis, acute necrotizing hemorrhagic encephalomyelitis, Guillain-Barré syndrome

    B.   **Dysmyelinating disease (leukodystrophy)**

        1.   **Defective formation of myelin** or defective turnover resulting in **accumulation of abnormal myelin breakdown products**

        2.   Results from genetic enzymatic abnormality (**inborn error of metabolism**)

        3.   Metabolic disorders — **metachromatic leukodystrophy, Krabbe's disease** (globoid cell leukodystrophy), sudanophilic leukodystrophy, Pelizaeus-Merzbacher disease, **adrenoleukodystrophy**, Canavan's disease (spongy degeneration; van Bogaert-Bertrand disease)

C. **Secondary demyelination (Wallerian degeneration)**

    1. **Axonal disruption** results in **degeneration of both distal axonal segment and its associated myelin sheath,** since integrity of myelin sheath requires continued contact with healthy viable axon

    2. **Axonal debris and myelin debris** (including neutral fats) identifiable in macrophages

II. **Multiple sclerosis (MS)**

A. Demyelinating disease characterized by **relapsing and remitting attacks** involving **multiple central nervous system locations** ("lesions disseminated in space and time")

B. No pathognomonic tests exist

    1. Clinical signs and symptoms relate to whatever nervous system structure is damaged during particular episode

    2. Clinical diagnosis requires correlation of history, examination findings, and laboratory studies to demonstrate pattern of relapsing and remitting neurologic signs and symptoms involving multiple central nervous system areas associated with laboratory findings consistent with demyelination

        a. **Magnetic resonance imaging (MRI)** can be used to identify areas of demyelination

        b. **Cortical evoked potentials** — abnormalities in visual, auditory, or somatosensory evoked potentials indicate lesions in myelinated fiber pathways

        c. **Cerebrospinal fluid (CSF)** examination — nonspecific findings of slightly increased cell count, protein, and total γ-globulin level, along with relatively specific findings of:

            (1) **Elevated IgG index** — elevated ratio of CSF to serum levels of IgG and albumin indicating increased synthesis of immunoglobulins inside blood-brain barrier

            (2) **Oligoclonal bands** — identification of several distinct bands of immunoglobulin using high-resolution gel electrophoresis of CSF

(3) **Elevated myelin basic protein level** — indicates active myelin breakdown

(4) Increased levels of free κ-light chains and increased numbers of T-lymphocytes

C. Pathologic features

1. Multiple irregular sharply circumscribed white matter **plaques**

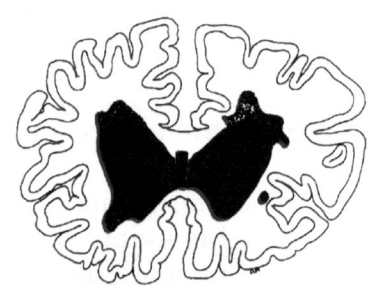

Periventricular multiple sclerosis plaques.

   a. Plaques are particularly prominent in periventricular cerebral white matter, optic nerves and tracts, basis pons, periventricular brain stem, and spinal cord

   b. Plaques particularly common at angle between caudate nucleus and corpus callosum

   c. Occasional plaques cross junction between cerebral cortex and white matter

2. Recent plaques tend to be grossly pink-tan and soft, while older plaques are grossly hyaline gray and rubbery or cystic

3. Microscopic appearance of plaques — complete or nearly complete **destruction of myelin** (with relative **preservation of axons**), proliferation of reactive astrocytes, and accumulation of macrophages (particularly in perivascular spaces)

4. **Shadow plaques** — area surrounding typical plaque in which myelin sheaths are thinner than normal (with consequent lighter staining in myelin stained slides) due to **remyelination**

D. Special forms

1. **Neuromyelitis optica (Devic's disease)**

   a. Relatively acute concurrent development of **visual symptoms (blindness)** and **spinal cord symptoms (paraplegia)**

        b.   Plaques identifiable in optic nerves or tracts and in spinal cord (often thoracic)

    2.   Acute multiple sclerosis (Marburg type)

        a.   Rare variant characterized by subacute neurologic deterioration with signs of cerebral, brain stem, spinal cord, and optic nerve dysfunction, progressing over several months to death; occasional brief remissions can occur during course

        b.   Large areas of demyelination along with intense lymphocytic and monocytic inflammation

        c.   Diffuse inflammation in acute multiple sclerosis distinguishes this disease from acute disseminated encephalomyelitis in which inflammation has perivenous distribution

E.   Etiology of multiple sclerosis is unknown; probably results from interaction between genetic propensity and environmental factors that alter immunologic functioning

F.   Treatment is unsatisfactory, but some benefit has been obtained with various immunosuppressive agents including corticosteroids

III.   Acute disseminated encephalomyelitis

A.   **Monophasic, self-limited demyelinating disease**; often fatal (in up to 50% of cases)

B.   Presumably results from immunologic attack directed against central nervous system myelin

    1.   **Postvaccinal** encephalomyelitis — occurs one to two weeks following vaccination against smallpox or rabies (particularly with brain-derived vaccine)

    2.   **Postinfectious** encephalomyelitis — occurs two to six days following onset of viral exanthem (such as measles, chickenpox, smallpox, or rubella) or rarely preceding exanthem

C.   **Clinical symptoms**

    1.   **Abrupt onset of headache, fever, confusion, and stiff neck**

2. Severe cases progress to convulsions, cerebellar ataxia, quadriplegia, cranial nerve palsies, or coma

3. **Acute cerebellitis (acute cerebellar ataxia)** form most common following chickenpox (varicella)

D. Pathologic features

1. **Symmetric involvement of entire neuraxis**

2. Intense **perivenous inflammation** consisting of lipid-laden macrophages and lymphocytes associated with zone of **perivenous demyelination** (and relative sparing of axons)

E. Treatment with immunosuppressive agents (including corticosteroids) can reduce disease severity

F. **Experimental allergic encephalomyelitis (EAE)**

1. Experimental model of immunologically-mediated central nervous system demyelinating disease

2. **Animals sensitized by injection of central nervous system myelin or myelin components** (such as myelin basic protein or proteolipid protein) dissolved in Freund's adjuvant (an immune system stimulant composed of paraffin oil, killed mycobacterium, and emulsifier) resulting in demyelinating illness

   a. Adult animals — monophasic inflammatory central nervous system demyelination (similar to human acute disseminated encephalomyelitis)

   b. Young animals — relapsing-remitting demyelinating illness (similar to human multiple sclerosis)

IV. **Acute necrotizing hemorrhagic encephalomyelitis**

A. **Acute, fulminant diffuse central nervous system demyelinating disease** progressing to coma and death within several days

B. Hyperacute form of acute disseminated encephalomyelitis

C. Pathologic features

1. Diffuse cerebral edema with petechial hemorrhages and large confluent areas of hemorrhagic necrosis involving white matter

2. **Fibrinoid necrosis of venules** with perivascular edema, fibrin deposition, neutrophilic infiltration, and necrosis

V. Guillain-Barré syndrome (inflammatory polyradiculoneuropathy)

A. Progressive, **symmetric, flaccid, areflexic motor weakness** often associated with antecedent viral illness, surgery, or immunization

B. Cerebrospinal fluid protein elevation with few cells (**albuminocytologic dissociation**)

C. Physiologic evidence of peripheral nerve demyelination — prolonged F-wave latency, slowing of nerve conduction velocities, and absent H-reflex

D. Pathologic features in peripheral nerve

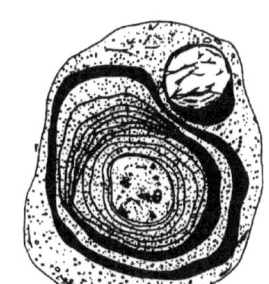

Guillain-Barré syndrome: intact central axon surrounded by macrophage cytoplasm digesting delaminated myelin.

1. **Perivascular inflammation** consisting of lymphocytes and macrophages

2. **Segmental demyelination** — macrophages penetrate between myelin lamellae and digest myelin sheath resulting in naked axonal segment covered only by thin rim of remnant Schwann cell cytoplasm

3. With recovery, **remyelination** occurs in demyelinated segments; remyelination results in shorter internode segments with thinner myelin sheath

E. Usually occurs as **monophasic illness**

F. **Relapsing polyradiculoneuropathy** (inflammatory demyelinating polyneuropathy, IDPN) — chronic disorder associated with repeated episodes of demyelination and remyelination

G. **Experimental allergic neuritis (EAN)** — experimental model similar to experimental allergic encephalomyelitis, except that **peripheral nerve myelin** is used to sensitize animals

SUGGESTED ADDITIONAL READING

McFarlin DE, McFarland HF: Multiple sclerosis. *N Engl J Med* 1982;307:1183-1188 & 1246-1251.

Raine CS: Demyelinating diseases. In Davis RL, Robertson DM (eds): *Textbook of Neuropathology*. Second Edition. Baltimore, Williams & Wilkins, 1991. pp. 535-620.

Vinken PJ, Bruyn GW, Klawans HL, Koetsier JC (eds): *Demyelinating Diseases. Handbook of Clinical Neurology. Volume 47*. Amsterdam, Elsevier Science Publishers, 1985.

CHAPTER 13: NERVOUS SYSTEM MALFORMATIONS

I. **Definitions**

　　A. **Malformation**

　　　　1. Morphologic defect of organ, part of organ, or larger region of body resulting from intrinsically abnormal developmental process

　　　　2. Most malformations are due to field defects

　　　　　　a. **Field defect** — abnormality of morphogenic field

　　　　　　b. Morphogenic field — region or part of embryo which responds as coordinated unit to embryonic interaction resulting in complex or multiple anatomic structures

　　B. **Disruption** — morphologic defect resulting from interference with originally normal development

　　C. **Deformation** — abnormal form or shape caused by mechanical forces

　　D. **Dysplasia** — abnormal organization of cells into tissues

　　E. **Dysraphia** — defective closure of the dorsal components of spine; includes spina bifida and related anomalies

II. **Neural tube defects (dysraphic disorders)**

　　A. **Anencephaly**

　　　　1. **Lethal anomaly** occurring in about 1 out of every 1000 live births

　　　　2. Often associated with **polyhydramnios**

3. Primary event is failure of anterior neuropore closure at about the fourth week of gestation

4. **Multifactorial inheritance** with recurrence risk approaching 5%

5. Pathologic features

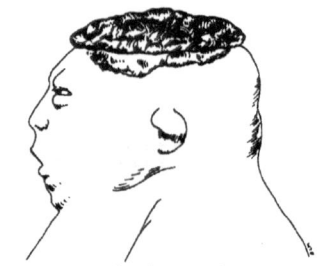

Anencephalic infant with bulging eyes and area cerebrovasculosa

   a. **Absence** or hypoplasia **of calvaria** (convexity skull bones)

   b. **Area cerebrovasculosa** — irregular masses of glial tissue, ependyma, and choroid plexus surrounded by heavily vascularized loose meningeal tissue

   c. Brain stem — identifiable but nuclear groupings often distorted; cranial nerves III through XII and ganglia are usually normal; optic nerves end blindly in orbit

   d. Cerebellum — disorganized cerebellar cortex identifiable

   e. Pituitary gland — normal anterior lobe, but absent posterior lobe; anterior lobe cells contain hormones, but are non-secretory

   f. Adrenal glands — hypoplastic

   g. Vertebrae

      (1) Bony defect may also involve dorsal arches of vertebral column

      (2) **Craniorachischisis — absence of all dorsal vertebral arches** in association with anencephaly

6. Associated malformations

   a. Cleft lip and palate

   b. Cyclopia

   c. Syndactyly

   d. Renal and cardiac anomalies

7. Detection possible in first trimester with finding of elevated serum or amniotic fluid α-fetoprotein levels, elevated amniotic fluid acetylcholinesterase levels, or abnormal cranium by ultrasonography

8. **Amnion disruption sequence**

   a. Sporadic event due to rupture of amnion resulting in interruption of normal morphogenesis, **deformation** of established structures, and mutilation of existing tissues

   b. Must be distinguished from primary anencephaly

B. **Spina bifida**

   1. **Spina bifida occulta** — open vertebral arches; often found incidentally on radiologic examination; skin anomalies include dimple, sinus, hypertrichosis, lipoma, or vascular nevus

   2. **Spina bifida cystica** — vertebral arch defect associated with cystic lesion

   3. **Meningocele**

      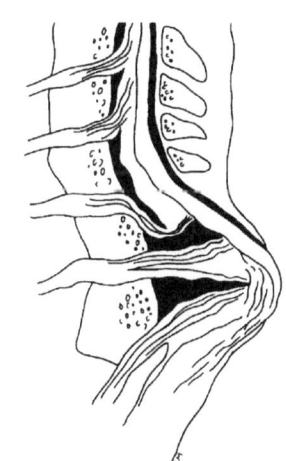

      a. Herniation of only dura and arachnoid through vertebral defect, with spinal cord left in normal position

      b. Most common site is lumbosacral spine

      c. No neurologic deficit

      Myelomeningocele.

   4. **Myelomeningocele (meningomyelocele)**

      a. Both meninges and spinal cord herniate through large vertebral defect

      b. Open spinal canal contains vascular mass consisting of remnants of gliotic neural tissue centrally with atrophic skin at margins; nerve roots are present to varying degrees

      c. Most common site is lumbosacral cord; less frequent at higher levels

d.   Neurologic abnormality is common; higher defects associated with greater neurologic loss

e.   **Arnold-Chiari (Chiari type II) malformation**

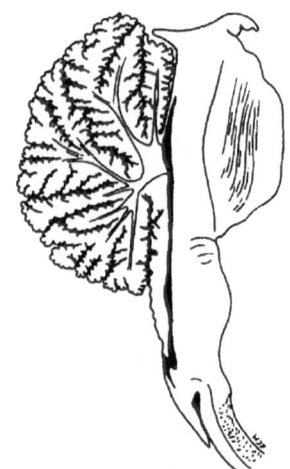

Arnold-Chiari malformation.

(1)   **Brain malformation** associated with myelomeningocele

(2)   Major components

(a)   **Elongated unrolled cerebellar vermis** displaced downward through foramen magnum

(b)   **Elongated kinked medulla** extending downward through foramen magnum

(3)   Associated anomalies

(a)   **Beaked collicular plate** — fused colliculi producing pointed midbrain tectum

(b)   **Aqueductal stenosis**

(c)   Subependymal heterotopic islands of gray matter

(d)   Bony skull anomalies

f.   Chiari type I malformation — cerebellar tonsillar herniation in adults; usually not associated with dysraphic state

III.   **Trisomy 21 syndrome (Down's syndrome)**

A.   Most **common** cause of mental retardation (occurring in up to 1 out of every 600 live births and accounting for 15% of all institutionalized patients)

B.   Increasing incidence with increased maternal age; 95% result from chromosomal nondisjunction

C. Clinical features include hypotonia, poor Moro reflex, hyperextensibility of joints, loose skin of posterior neck, flat facial profile, upslanting palpebral fissures, short ears, dysplastic pelvis, clinodactyly of fifth finger, simian palmar creases

D. Associated anomalies

1. Cardiovascular — atrioventricular canal defect, ventricular septal defect, patent ductus arteriosus, atrial septal defect, aberrant subclavian artery

2. Gastrointestinal — tracheoesophageal fistula, pyloric stenosis, duodenal atresia, annular pancreas, Hirschsprung's disease, imperforate anus

E. Pathologic features

1. Growth delay, microcephaly, **foreshortened frontal lobes**, low brain weight, narrow superior temporal gyrus

2. Histologic **evidence of Alzheimer's disease with increasing age** (invariably present over age 30 years)

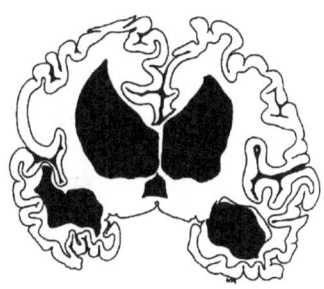

Hydrocephalus.

IV. **Hydrocephalus (hydrocephaly)**

A. **Abnormal enlargement of ventricular cavities due to accumulation of cerebrospinal fluid (CSF)**

1. **Communicating hydrocephalus** — enlarged ventricles communicate with subarachnoid space without obstruction

2. **Non-communicating (obstructive) hydrocephalus** — obstruction to cerebrospinal fluid flow preventing cerebrospinal fluid from entering subarachnoid space with resultant ventricular enlargement

B. Associated conditions include cranial enlargement, prominence of forehead, thinning of bones of cranial vault, enlargement of sutures and fontanels, craniofacial disproportion

C. **Obstetric hydrocephalus**

1. Hydrocephalus occurring in utero; often associated with polyhydramnios and breech presentation

2. Most such infants are stillborn or die in neonatal period; only 3% survive to leave hospital nursery

D. **Infantile and childhood hydrocephalus**

1. Hydrocephalus detected sometime after birth (up to age 20 years)

2. Presentation in early infancy usually associated with myelomeningocele (Arnold-Chiari malformation

3. Later presentation associated with other neurologic disorders including tumor, perinatal hemorrhage, meningitis, or aqueductal stenosis

E. **Acquired hydrocephalus** — associated with numerous conditions

1. **Oversecretion of cerebrospinal fluid — choroid plexus papilloma** (most papillomas obstruct cerebrospinal fluid flow and do not oversecrete cerebrospinal fluid)

2. **Neoplasms**

   a. Within cerebrospinal fluid pathways — colloid cyst, third ventricular tumors (such as germinoma or teratoma), astrocytomas of midbrain, medulloblastoma

   b. Outside cerebrospinal fluid pathways, but compressing aqueduct or third ventricle — craniopharyngioma, pituitary adenoma, pineal region tumors

   c. Within subarachnoid space and obstructing flow or egress of cerebrospinal fluid from arachnoid granulations — epidermoid tumors, lymphomas or leukemia, meningeal carcinomatosis

3. Obliteration of subarachnoid cerebrospinal fluid flow (typically at incisura or Pacchionian granulations)

   a. Subarachnoid blood (either primary hemorrhage or hemorrhage secondary to trauma)

   b. Inflammation, such as with bacterial or fungal meningitis

4. Metabolic disorders

   a. Mucopolysaccharidoses

    b.   Mannosidosis

F. **Normal pressure hydrocephalus**

    1.   Elderly individuals developing triad of **progressive dementia, urinary incontinence, ataxia**

    2.   CSF pressure may be normal, low, or elevated

    3.   Etiology unknown, but associated with conditions affecting CSF flow

G. **Hydrocephalus ex vacuo**

    1.   **Most common form** of hydrocephalus

    2.   **Associated with any condition which reduces brain parenchymal mass,** including stroke and neurodegenerative dementias (such as Alzheimer's disease)

    3.   Ventricular enlargement may be generalized or focal depending on site and extent of parenchymal loss

H. **Genetic hydrocephalus**

    1.   Rare; typically inherited as sex-linked recessive trait

    2.   Pathologic features vary but include aqueductal stenosis, communicating hydrocephalus, or Dandy-Walker malformation

I. **Dandy-Walker malformation**

    1.   **Cyst-like enlargement of fourth ventricle**

    2.   Enlargement of posterior fossa with elevation of transverse sinuses and tentorium

    3.   Lack of patency of fourth ventricular outflow foramina (foramen of Magendie and foramina of Luschka)

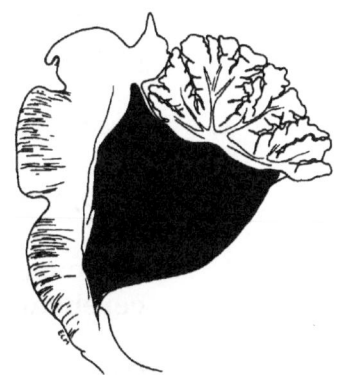

Dandy-Walker malformation.

    4.   Associated with profound mental retardation

V.   **Anomalies of neuronal migration**

   A.   Normal cerebral cortical development

   1.   **Germinal matrix** — periventricular germinal zone that produces neuroblasts

   2.   **"Inside-out development"** — cerebral cortical layers normally develop by successive waves of neuroblast migration from germinal matrix, with each successive wave of migrating neuroblasts passing earlier waves to form more superficial layers

   3.   Interference with normal neuronal migration produces abnormal cerebral cortex and white matter

   B.   **Pachygyria and agyria**

   1.   **Pachygyria** — reduced number of coarse cerebral gyri

   2.   **Agyria**

   a.   **Smooth brain with absent cerebral gyri**

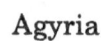

Agyria.

   b.   **Lissencephaly syndrome** (Miller-Dieker syndrome) — severe microcephaly, furrowed forehead, neonatal jaundice, purpura, seizures, and agyria

   3.   Cerebral cortex composed of four or fewer layers (instead of normal six layers)

   4.   Associated with many conditions including chromosomal abnormalities

   C.   **Polymicrogyria**

   1.   **Increased number of small shallow cerebral gyri**

   2.   Cerebral cortex composed of four or fewer layers with fusion of adjacent gyri

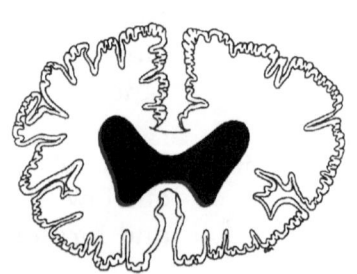

   3.   May be focal, unilateral, or involve entire cortex of both hemispheres

Polymicrogyria.

4.   Associated with microcephaly

5.   Associated anomalies include polymicrogyria of cerebellum, hyperconvoluted dentate nucleus, or hypoplastic corticospinal tract

VI.   Hydranencephaly

A.   **Absence of cerebral hemispheres** typically in distribution of anterior and middle cerebral arteries with **intact meninges** and normal skull

B.   Pathogenesis

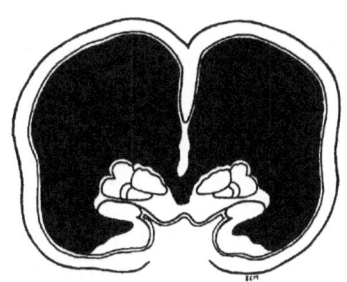

1.   **Acquired destruction of brain** (after formation of cerebral hemispheres, but long before birth) usually due to **extensive prenatal ischemic injury,** such as **bilateral carotid artery occlusion**

Hydranencephaly: enlarged CSF space with remnant basal ganglia, thalamus, and medial-inferior temporal lobe.

2.   **Remnant brain tissue mainly in posterior fossa, thalamus, and basal ganglia (vertebrobasilar artery distribution)**

3.   Etiology unknown but has been associated with prenatal infections (toxoplasmosis, fungal, viral) or maternal trauma

4.   Clinical findings are nonspecific including seizures, spasticity, and developmental delay (mental retardation)

5.   Prolonged survival can occur, particularly if hypothalamus is intact

VII.   Porencephaly

A.   **Cavity or cavities within cerebral hemispheres,** often bilateral and symmetric

B.   Term should be restricted to large circumscribed defects in cerebral hemisphere

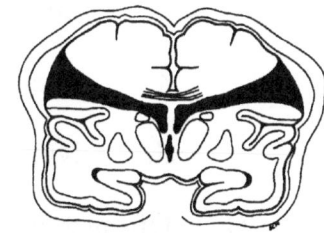

C.   Prevalence is unknown; many cases discovered incidentally in adults

Bilateral porencephalic cysts.

D.   Possible etiologies include trauma, circulatory disturbance, or infections

E.   Clinical findings are nonspecific including spasticity, seizures and developmental delay or mental retardation

F.   Associated anomalies include microcephaly, agenesis of olfactory lobes, congenital cataracts, bifid uvula, or malformed ears

VIII.   Multicystic encephalopathy

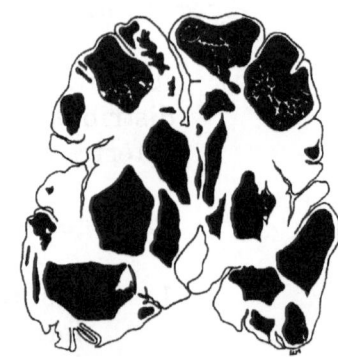

A.   Numerous cavities in brain involving both gray and white matter due to infarction of brain tissue

B.   Etiology unknown but damage occurs late in utero from circulatory disturbance

C.   Differs from porencephaly and hydranencephaly only in degree and extent of tissue necrosis

Multicystic encephalopathy.

IX.   Holoprosencephaly

A.   Complex malformation with impairment of midline cleavage of embryonic forebrain (**field defect**) associated with various gradations of facial dysmorphic features

B.   Pathologic features

    1.   **Single large ventricle** with fused anterior cerebrum

    2.   Large posterior and dorsal cyst in some cases

    3.   **Absence of olfactory tracts and nerves (arrhinencephaly)**

    4.   Fusion of thalami

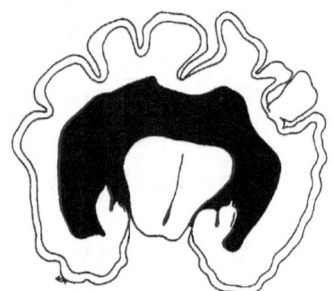

    5.   Numerous subarachnoid glial heterotopias

Holoprosencephaly.

C.   Often associated with chromosomal anomalies, particularly **trisomy 13 (Patau's syndrome)** or less commonly trisomy 18 (Edward's syndrome)

D.   Degree of mental retardation dependent upon severity of brain anomaly

E.   Associated facial anomalies include cyclopia, presence of proboscis, flattened nasal area (cebocephaly), or median cleft lip

X.   **Encephalocele**

A.   **Herniation of cerebral tissue** through defect in cranium (occipital, parietal, frontal, or basal portions)

B.   **Occipital encephalocele**

1.   Variable size of bony defect in occipital region and amount of included cerebral tissue

2.   Can be asymptomatic or associated with hydrocephalus, ocular palsies, or visual disturbance

3.   Associated malformations include syndactyly, lung anomalies, omphalocele

4.   **Meckel's syndrome** — occipital encephalocele associated with cleft palate, polydactyly, ambiguous genitalia, polycystic kidneys, congenital heart disease, eye defects

C.   **Parietal encephalocele**

1.   Interparietal midline skull defect

2.   Associated malformations include absent corpus callosum and Dandy-Walker malformation

D.   Anterior sincipital and basal encephaloceles

1.   Bony defect in skull base

a.   Sincipital — visible

b.   Basal — not visible

2.   Associated features include hypertelorism, exophthalmos, abducens nerve palsy, syndactyly, and optic nerve anomalies

XI.   **Agenesis of the corpus callosum**

A.   **Absence or partial absence of corpus callosum**

B.   Prevalence is unknown, but often found incidentally in adults

C.   Can be asymptomatic or associated with mental retardation when present with other brain anomalies

D.   Etiology unknown; some cases are familial (recessive or sex-linked inheritance) or associated with chromosomal abnormalities

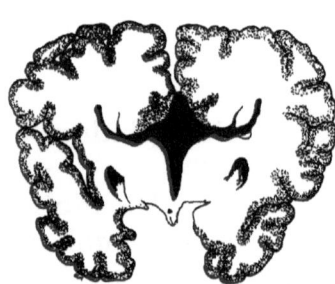

Agenesis of corpus callosum.

E.   Pathologic features

1.   Interhemispheric surface of cerebral hemisphere has **no cingulate gyrus** and gyri radiate perpendicular to roof of third ventricle

2.   Coronal sections of cerebral hemispheres show **upturned pointed corners of lateral ventricles ("butterfly ventricles")**

3.   **Probst bundles** — stumps of white matter at interhemispheric edge of cerebral cortex

F.   Associated syndromes

1.   **Aicardi syndrome** — infantile spasms, mental retardation, chorioretinopathy, vertebral anomalies

2.   **Shapiro syndrome** — recurrent hypothermia

3.   **Andermann syndrome** — mental retardation and sensorimotor neuropathy with progressive flaccid paraplegia

XII.   **Arrhinencephaly**

A.   Absence of olfactory tract

B.   May be isolated finding or in association with other anomalies such as holoprosencephaly

C.   **Kallmann's syndrome** — sex-linked recessive disorder characterized by arrhinencephaly

SUGGESTED ADDITIONAL READING

DeMyer W, Zeman W: Alobar holoprosencephaly (arrhinencephaly) with median cleft lip and palate: clinical, electroencephalographic and nosologic considerations. *Confin Neurol* 1963; 23:1-36.

Frerebeaue P, Dimeglio A, Gras M, Harbi H: Diastematomyelia: Report of 21 cases surgically treated by a neurosurgical and orthopedic team. *Child's Brain* 1983; 10:328-339.

Friede RL: *Developmental Neuropathology*, Second revised and expanded edition. Berlin, Springer-Verlag, 1989.

Hori A, Fischer G, Dietrich-Schott B, Ikeda K: Dimyelia, diplomyelia, and diastematomyelia. *Clin Neuropathol* 1982; 1:23-30.

Leech RW, Shuman RM: Holoprosencephaly and related midline cerebral anomalies: a review. *J Child Neurol* 1986; 1:3-18.

Loeser JD, Alvord, Jr. EC: Clinicopathological correlations in agenesis of the corpus callosum. *Neurology* 1968; 18:745-756.

Opitz JM, Reynolds JF, Spano LM: *The Developmental Field Concept*. New York, Alan R. Liss, 1986.

Parrish ML, Roessmann U, Levinsohn MW: Agenesis of the corpus callosum: a study of the frequency of associated malformations. *Ann Neurol* 1979; 6:349-354.

Sarnat HB: *Cerebral Dysgenesis: Embryology and Clinical Expression*. New York, Oxford University Press, 1992.

Winter RM, Knowles SAS, Bieber FR, Barraitser M: *The Malformed Fetus and Stillbirth. A Diagnostic Approach*. Chichester, John Wiley & Sons, 1989.

CHAPTER 14: SPINAL CORD PATHOLOGY

I. **Vascular disorders**

    A. Ischemia or infarction — pathology is highly variable from complete segmental necrosis at multiple levels to necrosis of portions of anterior spinal cord at one level

        1. **Rostral** spinal cord

            a. Uncommon disorder produced by **occlusion of rostral portion of anterior spinal artery** which also supplies anterior and median portion of lower medulla

            b. Clinical symptoms — pain in back and neck, paralysis of both arms, loss of pain and temperature sensation, preservation of proprioception

        2. **Mid-thoracic** spinal cord

            a. Produced by **occlusion of caudal portion of anterior spinal artery usually secondary to occlusion of feeding radicular arteries** branching from aorta, as occurs in dissecting aortic aneurysm or abdominal aortic aneurysm

            b. Clinical symptoms — back pain, paraplegia, loss of sensation below level of lesion (sensory level)

    B. **Arteriovenous malformation**

        1. **Large tortuous thick walled vessels on posterior surface of spinal cord** with numerous small thick walled vessels within substance of spinal cord (demonstrable both radiologically and pathologically)

        2. Incomplete necrosis of both gray and white matter with reactive gliosis

II. **Inflammatory conditions**

A. **Varicella-zoster virus**

1. Clinical symptoms — **skin vesicles in dermatome of** a sensory nerve and radicular pain or altered sensation

2. Pathologic features — intense mononuclear cell inflammation and thrombosed vessels typically **limited to dorsal root ganglia** (usually only one ganglion involved) but can extend along sensory root to involve spinal cord

B. **Poliomyelitis**

1. Clinical symptoms — acute onset of asymmetric flaccid paralysis following brief febrile illness

2. Pathologic features

    a. Early — **hemorrhagic necrosis of anterior horn gray matter, lymphocytic infiltration** of meninges and anterior horn, and neuronophagia with microglial nodules (neuronal tombstones)

    b. Late — loss of anterior horn neurons and myelinated fibers in anterior spinal nerve root; neurogenic atrophy of muscle

C. **Epidural abscess**

1. Clinical symptoms — localized back pain and tenderness and paraparesis

2. Produced by spread of infection from **adjacent spinal osteomyelitis**; common organisms include staphylococcus, *Streptococcus pneumoniae*, or *Pseudomonas aeruginosa*

3. Pathologic features — localized or diffuse inflammation, thrombosis of blood vessels, and compression of spinal cord

D. **Spinal tuberculosis** (*Mycobacterium tuberculosis* infection)

1. **Tuberculous osteitis** — vertebral tuberculosis resulting in bony destruction and collapse with consequent spinal column deformity and compression of spinal cord; inflammation can spread through dura to produce tuberculous meningitis

2. **Tuberculous meningitis** — tuberculous infection of subarachnoid space with granulomas and fibrosis; resultant arteritis can produce spinal cord infarcts

E. **Tabes dorsalis**

1. Form of **neurosyphilis** developing approximately 15-20 years after initial infection

2. Clinical symptoms

   a. Sharp stabbing pains (**lightning pains**) in lower extremities, progressive loss of proprioception with consequent sensory **ataxia**, and loss of pain sensibility

   b. **Charcot joints** — destruction of proximal joints (such as hips and knees) which are repeatedly injured by abnormal positioning (from loss of proprioception) and have reduced pain sensibility

3. Pathologic features — **destruction (with minimal inflammation) and fibrosis of dorsal spinal roots** (particularly thoracic and lumbar) with **degeneration of spinal cord dorsal columns** resulting in atrophic and dorsally-flattened spinal cord; **dorsal root ganglia are not involved**

F. **Human immunodeficiency virus-1 (HIV-1)-associated myelopathy**

1. Clinical symptoms — progressive paraparesis, ataxia, and incontinence

2. Pathologic features — **vacuolar changes in spinal cord dorsal and lateral columns** associated with gliosis and occasional multinucleated giant cells

G. **Tropical spastic paraparesis** (chronic progressive myelopathy)

1. Myelopathy from **human T-cell lymphotropic virus-I (HTLV-I)** infection; endemic in Caribbean islands and other tropical regions as well as in Japan

2. Clinical symptoms — **progressive spastic paraparesis** (leg weakness), loss of bowel and bladder sphincter tone (incontinence), and minimal or no sensory abnormality; sometimes associated with T-cell lymphoblastic leukemia or lymphoma

3. Pathologic features — **lymphocytic infiltration of leptomeninges and spinal cord** (chronic meningomyelitis) with loss of myelinated fibers in lateral and anterior spinal cord columns

III. **Spinal trauma**

A. Indirect injury

1. Forces generated by rotation, sudden flexion, hyperextension, or vertebral compression

2. Associated with fracture of vertebral bodies, dislocation of facet joints, misalignment of spinal canal, herniation of intervertebral disc material, or splintering of bone

B. Direct injury — missile or sharp objects

C. Most common sites of traumatic lesions are **cervical cord** (particularly lower cervical, C5-T1) and **thoracolumbar junction** (T11, T12, L1)

D. Vertebral column damage

1. **Atlanto-occipital dislocation** — often immediately fatal due to vertebral artery tearing or medullary compression

2. **Jefferson fracture** — blow to vertex of head resulting in **split posterior arch of C1 (atlas)**; if not immediately fatal due to spinal cord concussion, can have no associated neurologic signs since vertebral canal is widened by fracture

3. **Odontoid injuries — separation of odontoid** can be fatal due to medullary compression, either immediately or following subsequent neck movement

4. **Hangman's fracture — disruption of arch of C2** along with **C2-C3 dislocation**; usually results from hyperextension

5. **Locked facets** — facet dislocation; unilaterally associated with root injury; bilaterally associated with spinal cord injury

6. Dislocations through disc space or vertebral body — marked misalignment with severe spinal cord injury

7. **Wedge fractures** — common in **thoracic** vertebral injuries

8. **Vertebral body bursting** — common with **thoracolumbar** vertebral injuries; results in **free floating bone fragments** in spinal canal pressing on spinal cord and roots

E. **Spinal cord concussion** — similar to brain concussion and characterized by temporary impairment of spinal cord function without evidence of vertebral fracture or dislocation

F. **Spinal cord contusion-laceration** — similar to brain contusion-laceration with impairment of spinal cord function due to **bruising and tearing** of tissue and resultant necrosis, hemorrhage, and edema

G. Pathologic features

1. Varies from superficial **contusion to transection**

2. Early changes relate to small hemorrhages, necrosis, and edema with subsequent resolution to scarring and formation of cysts or **syringomyelic cavities**

3. Spinal cord is often crushed with obliteration of subarachnoid and subdural spaces

4. Lesion usually **extends above and below primary site in cone-shaped distribution**

5. Histologic appearance

   a. Initially severe edema, petechiae, acute inflammation, disruption of myelin and axons, ischemic neurons

   b. Subsequent cavitation and gliosis extending one to two segments above and below primary site

IV. **Degeneration of intervertebral disc**

A. Progressive loss of water results in shrinkage and fibrosis of intervertebral disc

B. Weakening of annulus results in disc protrusion

C. **Osteophytes** — fibrosis (and subsequent mineralization) of protruding annulus; often impinge on spinal roots

V.   **Rheumatoid arthritis** — destruction of articular joint surfaces and capsules (particularly in cervical level) resulting in cervical vertebral dislocation

VI.   **Ossification of posterior longitudinal ligament** — results in spinal cord compression from spinal canal stenosis

VII.   **Syringomyelia**

    A.   Clinical symptoms — loss of pain and temperature sensation in arms, neck and arm pain, upper extremity muscle atrophy (particularly in hands)

    B.   Pathologic features — **cavity in central portion of spinal cord,** lined by astrocytes, and often extending over several segments or into lower brain stem (syringobulbia)

VIII.   **Neoplasms**

    A.   **Extradural neoplasms:** metastatic tumors such as lung cancer, breast cancer, lymphoma, multiple myeloma

    B.   **Intradural/extramedullary neoplasms** — meningioma, neurofibroma, schwannoma

    C.   **Intramedullary neoplasms** — ependymoma, astrocytoma

    D.   **Meningeal carcinomatosis** (subarachnoid neoplasms) — metastatic tumors such as lung cancer, breast cancer, gastric cancer

IX.   Spinal cord anomalies

    A.   **Hydromyelia** — focal or diffuse **enlargement of spinal cord central canal** with ependyma intact; most common in lumbar region

Normal spinal cord (A), extradural tumor (B), intradural/extramedullary tumor (C), and intramedullary tumor (D).

B. **Diastematomyelia — two hemicords** within single dural sac separated by vascular tissue or in two separate dural sacs with bony spur in between; typically associated with spina bifida

C. **Diplomyelia — duplicated spinal cord**; frequently associated with other anomalies of spinal cord and filum terminale

SUGGESTED ADDITIONAL READING

Byrne TN, Waxman SG: *Spinal Cord Compression: Diagnosis and Principles of Management.* Philadelphia, F. A. Davis Co, 1990.

Woolsey RM, Young RR (eds): Disorders of the spinal cord. *Neurol Clin* 1991; 9(3):503-816.

Vinken PJ, Bruyn GW, Braakman R (eds): *Injuries of the Spine and Spinal Cord-Part I. Handbook of Clinical Neurology. Volume 25.* Amsterdam, North-Holland Publishing Co, 1976.

Vinken PJ, Bruyn GW, Braakman R (eds): *Injuries of the Spine and Spinal Cord-Part II. Handbook of Clinical Neurology. Volume 26.* Amsterdam, North-Holland Publishing Co, 1976.

CHAPTER 15: PINEAL GLAND PATHOLOGY

I. Anatomy and physiology

A. **Pineal gland (epiphysis cerebri)** is situated in subarachnoid space behind third ventricle, below splenium of corpus callosum and origin of great vein of Galen, and above and between midbrain superior colliculi

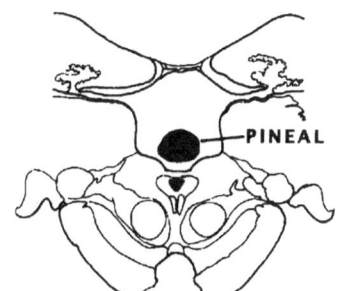

Pineal gland position in relation to midbrain aqueduct and tectum.

B. **Capillaries** supplying pineal gland parenchymal cells have flattened fenestrated endothelium and **no blood-brain barrier**

C. Pineal parenchyma contains **pineocytes** (pinealocytes) intermixed with astrocytes

D. **Melatonin** — present in **neurosecretory granules** in pineocytes and secreted into pineal capillaries

    a. Synthesized from **serotonin (5-hydroxytryptamine)** by **N-acetylation** (to form N-acetylserotonin) followed by O-methylation

Serotonin.

Melatonin.

     b.   Influences **sexual maturity (by inhibiting gonadal development** and delaying onset of puberty), reproductive activity, and thyroid and adrenal gland function

E. **Innervation** of pineal gland

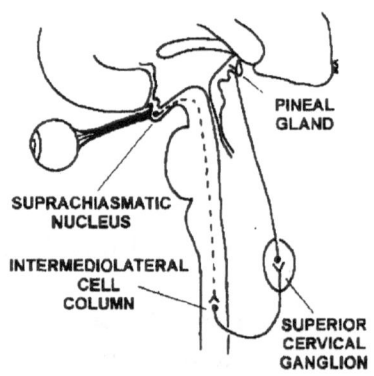

    1.   **Postganglionic sympathetic (noradrenergic) neurons** originating in **superior cervical ganglion** travel with pineal blood vessels to pineal gland

    2.   Pineal **light response**

Innervation of pineal gland.

        a.   Mediated by neural pathway involving **retinal input** to **suprachiasmatic nucleus of hypothalamus** which influences **spinal cord intermediolateral cell column neurons** that send axons to **superior cervical ganglion**

        b.   **Norepinephrine release** by pineal postganglionic sympathetic terminals:

            (1)   **Increases in darkness** and is suppressed by exposure of eyes to light

            (2)   **Stimulates pineocyte ß-adrenergic receptors** which increases levels of N-acetyltransferase (rate limiting enzymatic step in melatonin synthesis), thereby increasing melatonin levels

F. **Circumventricular secretory organs**

    1.   Family of specialized neural structures that includes **pineal gland,** subcommissural organ, area postrema, subfornical organ, median eminence (infundibulum), organum vasculosum of lamina terminalis, and neurohypophysis (posterior pituitary)

    2.   Embryonic derivation from modified ependymal cells and neurons

    3.   Vascular supply by **fenestrated capillaries** with **no blood-brain barrier; large molecules** (such as contrast-enhancing agents used in

radiographic studies) readily **enter interstitial space** of these structures

II. Pathology

A. **Pineal calcification** — calcification of ground substance secreted by pineocytes begins in childhood and increases with age; useful for identifying pineal region on radiographic studies

B. **Pineal cysts** — small cysts lined by proliferation of **astrocytes with prominent Rosenthal fibers**

C. **Pineal neoplasms**

1. Signs and symptoms

a. Pressure on midbrain produces **aqueductal obstruction** with resultant **obstructive hydrocephalus** of lateral and third ventricles and **increased intracranial pressure**

b. **Parinaud's syndrome** (dorsal midbrain syndrome)

(1) **Pressure on midbrain pretectum, superior colliculi, and posterior commissure**

(2) **Impairment of upgaze** and varying degrees of lid retraction, loss of pupillary light reflex (with preserved accommodation), and **downward gaze preference ("setting sun sign")**

c. **Anterior extension into third ventricle** with pressure on thalamus and hypothalamus results in mental changes and endocrine disturbances

d. Pineal destruction results in alteration of timing of sexual maturity (**precocious puberty** or **delayed puberty**)

e. Extension of tumor into third ventricle disrupts hypothalamic regulatory influences on gonadotropin release

f. **Subarachnoid dissemination** with distant symptoms

2. **Germ cell tumors** — histologically similar to germ cell tumors found in testes, ovary, mediastinum, or elsewhere

    a. **Germinoma** — intermixture of small lymphocytes and large undifferentiated cells with large nuclei and large nucleoli; similar to testicular seminoma and ovarian dysgerminoma; highly **radiosensitive** tumor

    b. **Embryonal carcinoma** — sheets of large undifferentiated cells; secretes **α-fetoprotein** identifiable in cerebrospinal fluid and blood

    c. **Choriocarcinoma** — hemorrhagic tumor mimicking **trophoblastic** tissue; produces **human chorionic gonadotrophin (HCG)** detectable in cerebrospinal fluid and blood

    d. **Teratoma**

        (1) Mature teratoma — slow growing spherical cystic mass containing wide-range of well-differentiated tissues including epithelium, hair, glands, cartilage, muscle, bone, or neural tissue; complete resection is curative

        (2) Immature teratoma — tumor containing mixture of multiple well-differentiated and poorly differentiated (often malignant) tissues

3. **Tumors of pineocytes** (pineal parenchymal cells)

    a. **Pineoblastoma — primitive neuroectodermal tumor (PNET)** originating in pineal gland

        (1) Composed of small primitive cells with high nuclear to cytoplasmic ratio and frequent rosette formation (and other features similar to retinoblastoma)

        (2) Often familial and associated with retinoblastoma ("trilateral retinoblastoma")

        (3) Poor prognosis because of rapid growth and subarachnoid spread

221

      b.   **Pineocytoma** — similar to pineoblastoma, but with more differentiated cells (resembling pineocytes) containing neurosecretory granules

4.   **Glioma** — similar to astrocytic neoplasms originating in other areas of cerebrum

5.   **Secondary tumors** — rich vascular supply and lack of blood brain barrier predispose to **metastases** to pineal gland

SUGGESTED ADDITIONAL READING

Bruce JN, Stein BM: Pineal tumors. *Neurosurg Clin North Am* 1990; 1:123-138.

Erlich SS, Apuzzo MLJ: The pineal gland: anatomy, physiology, and clinical significance. *J Neurosurg* 1985; 63:321-341.

Lewis PD: Pineal gland. In Weller WO (ed): *Systemic Pathology. Third Edition. Volume 4: Nervous System, Muscle and Eyes.* Edinburgh: Churchill Livingstone, 1990, pp. 504-515.

Martin JB, Reichlin S: *Clinical Neuroendocrinology.* Edition 2. Philadelphia, F. A. Davis Co., 1987.

Min K-W: Ontogeny of human pineal gland. *J Child Neurol* 1988; 3:275.

Wurtmann RJ, Moskowitz MA: The pineal organ. *N Engl J Med* 1977; 296:1329-1333 & 1383-1386.

CHAPTER 16: PITUITARY GLAND PATHOLOGY

I. Anatomy and physiology

A. Pituitary gland is located in **sella turcica** beneath **dural fold** (**diaphragma sellae**) and attached to hypothalamus by pituitary stalk

   1. **Floor of sella turcica** forms portion of roof of **sphenoid sinus**

   2. **Lateral wall of sella turcica** is medial wall of **cavernous sinus** which contains (intracavernous portions) internal carotid artery and cranial nerves III (oculomotor nerve), IV (trochlear nerve), V (trigeminal nerve), and VI (abducens nerve)

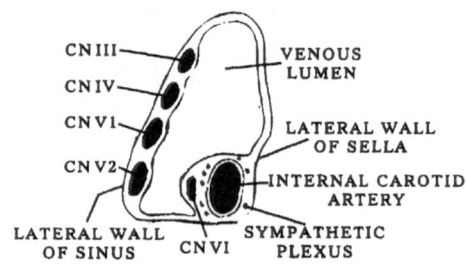

Structures within cavernous sinus adjacent to sella turcica.

B. Embryonic derivation

   1. **Adenohypophysis** — glandular structure derived from evagination of roof of posterior nasopharynx (**Rathke's pouch**)

      a. Anterior portion of Rathke's pouch gives rise to anterior lobe

      b. Posterior portion (vestigial lumen) of Rathke's pouch gives rise to thin intermediate lobe which is hormonally inactive postnatally but contains colloid-filled cysts with ciliated lining cells

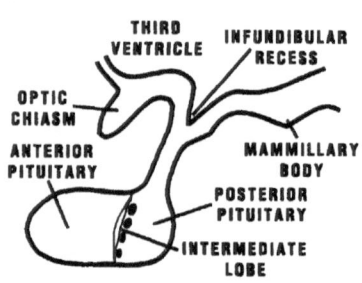

Pituitary gland.

2. **Neurohypophysis (posterior pituitary)** — neural bud from diencephalic floor grows downward to abut posterior portion of Rathke's pouch

C. Vascular supply

   1. Neurohypophysis — arterial supply from inferior hypophyseal artery (branch of cavernous portion of internal carotid artery)

   2. Adenohypophysis — portal blood supply from hypophyseal portal system

      a. Superior hypophyseal arteries (branches of intracranial internal carotid artery) form arterial ring around hypothalamic infundibulum (median eminence) and pituitary stalk providing vascular supply of fenestrated capillaries to these structures

      b. Various hypothalamic neurons send axons in tuberoinfundibular tract where hypophysiotrophic factors (anterior pituitary releasing and inhibiting hormones) are released into fenestrated capillaries

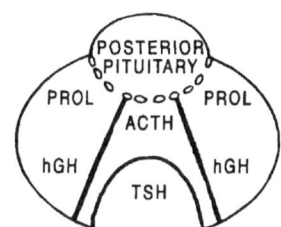

Distribution of pituitary hormones in horizontal section of pituitary gland.

      c. Infundibular capillaries drain into portal veins that course into anterior lobe capillary network where hypophysiotrophic factors influence hormone secretion by anterior pituitary cells

D. Cellular architecture of adenohypophysis — regional cell distribution except for gonadotrophs which are diffusely distributed

   1. **Corticotrophs**

      a. Large densely granulated **basophilic** cells producing **ACTH** and other peptides derived from **proopiomelanocortin** (POMC)

      b. **Crooke's hyaline change** — perinuclear eosinophilic hyaline ring resulting from massive accumulation of normal perinuclear cytokeratin intermediate filaments under conditions of glucocorticoid excess (endogenous or exogenous)

   2. **Thyrotrophs** — moderately granulated **basophilic** cells producing **TSH**

3. **Gonadotrophs** — moderately granulated **basophilic** cells producing **LH, FSH,** or both hormones

4. **Somatotrophs** — densely granulated **eosinophilic** cells producing human growth hormone

5. **Lactotrophs**

    a. Densely granulated (**eosinophilic**) or sparsely granulated (**chromophobic**) cells: eosinophilic cells represent storage phase, while chromophobic cells are actively secreting prolactin

    b. Mammosomatotroph cells contains both prolactin and human growth hormone

    c. **Hypertrophy and hyperplasia of chromophobic lactotrophs** during **pregnancy and lactation ("pregnancy cells")** results in doubling of pituitary volume; similar hypertrophy can be seen with estrogen administration (in birth control pills)

E. **Neurohypophyseal hormones**

1. Produced by large (magnocellular) **hypothalamic neurons** located in **supraoptic nuclei** (above optic tracts) and **paraventricular nuclei** (adjacent to third ventricle)

    a. Nonapeptide (nine amino acid polypeptide) hormones:

        (1) **Vasopressin (antidiuretic hormone; ADH)** — primarily synthesized in supraoptic nucleus; **increases water permeability of renal collecting ducts** resulting in **concentrated urine** in order to conserve body water

        (2) **Oxytocin** — produced mostly by cells in paraventricular nucleus; **stimulates progression of labor (uterine contractions)** and **"milk let-down"** (contraction of breast myoepithelial cells)

    b. Cell bodies synthesize prohormones which are packaged in membrane-bound vesicles (neurosecretory granules), transported by axoplasmic flow to posterior pituitary, and cleaved into oxytocin (and neurophysin I) or vasopressin (and neurophysin II) in granules prior to release into fenestrated capillaries

2.   Pathophysiology

     a.   **Diabetes insipidus** — excretion of **large quantities of dilute urine**; usually associated with **increased water intake (polydipsia)**

         (1)   **ADH deficiency** — due to damage to pituitary stalk from trauma, surgical manipulation, granulomatous disease (particularly sarcoidosis), histiocytosis X (eosinophilic granuloma), or metastatic neoplasms

         (2)   **Nephrogenic diabetes insipidus — normal ADH secretion but lack of response by target renal tissues**; may be hereditary due to receptor defect or acquired due to pharmacologic blockade of ADH effect (as follows lithium administration for treatment of manic-depressive disease)

     b.   **Syndrome of inappropriate ADH secretion (SIADH)**

         (1)   **Excessive secretion of ADH in relation to degree of plasma hypoosmolality** resulting in progressive **dilutional hyponatremia** from excessive water retention and excretion of relatively concentrated urine

                 (a)   Neurogenic — various cerebral disorders (associated with cerebral edema) including trauma, tumor, infection, or infarction; usually resolves spontaneously over time

                 (b)   Ectopic production of ADH by tumors (particularly small cell anaplastic or oat cell lung cancer); treatment necessitates anticancer therapy plus drugs that block renal ADH effect

         (2)   Symptoms relate to **hyponatremia**: generally beginning as **lethargy and irritability** when serum sodium levels reach 120 mEq/L, and progressing to **stupor, coma, and seizures** as serum sodium levels reach 110 mEq/L

                 (a)   **Treatment** of hyponatremia involves **water restriction**

      (b)   Rapid correction with hypertonic saline infusions and diuretics (such as furosemide) can result in central pontine myelinolysis

F.   **Hypothalamic hypophysiotrophic hormones** — factors produced by small (parvicellular) neurons throughout hypothalamus

    1.   **Corticotropin releasing hormone (CRH)** — 41 amino acid polypeptide that **stimulates adrenocorticotropic hormone release**

    2.   **Thyrotropin releasing hormone (TRH)** — tripeptide (3 amino acid polypeptide) that **stimulates thyrotropin and prolactin release**

    3.   **Gonadotropin releasing hormone (GnRH)** — decapeptide (10 amino acid polypeptide) that **stimulates release of luteinizing hormone** and **follicle-stimulating hormone**

    4.   **Growth hormone releasing hormone (GHRH)** — 40 amino acid polypeptide that **stimulates growth hormone release**

    5.   **Somatostatin** — 14 amino acid polypeptide that **inhibits growth hormone and thyrotropin release**

    6.   **Dopamine** — **catecholamine** that **inhibits prolactin release**

    7.   **Vasoactive intestinal polypeptide (VIP)** — 28 amino acid polypeptide that **stimulates prolactin release**

G.   **Adenohypophyseal hormones** — anterior pituitary polypeptide hormones with multiplicity of systemic effects

    1.   **Adrenocorticotropic hormone (ACTH)**

        a.   **Proopiomelanocortin (POMC)** — glycosylated (~285 amino acid) **prohormone** present in brain and pituitary is cleaved into three general peptide groups:

            (1)   N-terminal fragment — inactive except for further cleavage into γ-melanocyte-stimulating hormone (γ-MSH)

            (2)   **ACTH** — 39 amino acid peptide ($ACTH_{1-39}$) [although synthetic N-terminal peptide $ACTH_{1-24}$ has same degree of clinical corticotropic activity]; can be further cleaved in intermediate lobe (prenatally) to form α-MSH ($ACTH_{1-13}$)

and corticotropin-like-intermediate lobe peptide (CLIP; $ACTH_{18-39}$)

(3) **ß-Lipotropin (ß-LPH)** — rapidly cleaved into γ-LPH and **ß-endorphin**; **α-endorphin** and **γ-endorphin** are derived by further cleavage of ß-endorphin

b. **Stimulates release of glucocorticoid hormone cortisol (hydrocortisone)**, along with sex steroid hormones androsterone and androstenedione from **adrenal gland**; necessary for continued **aldosterone** (mineralocorticoid) production, although secretion of aldosterone is regulated by renin-angiotensin system

c. Pathophysiology

(1) **Cushing's disease**

(a) **Bilateral adrenal hyperplasia** due to excess secretion of pituitary ACTH by pituitary corticotroph adenoma

(b) Must be differentiated from **Cushing's syndrome** which is due to **excess glucocorticoids** from various causes including exogenous administration, adrenal tumor, or adrenal stimulation from ectopic ACTH secretion by tumors (particularly small cell anaplastic or oat cell lung cancer)

(c) Clinical symptoms — **hyperglycemia**, muscle wasting, **osteoporosis**, negative nitrogen balance, **truncal obesity with fat accumulation over back between shoulders ("buffalo hump")**, and involution of lymphoid tissue with consequent increased susceptibility to infection

(2) **Nelson's syndrome** — skin hyperpigmentation and visual field loss following bilateral adrenalectomy for hypercortisolism; represents missed diagnosis of Cushing's disease with resultant aggressive (or malignant) growth of remaining pituitary corticotroph adenoma

2. **Thyrotropin (thyroid-stimulating hormone; TSH)**

    a.   **Glycoprotein** dimer composed of α-subunit (similar to α-subunit of FSH and LH) and ß-subunit

    b.   **Stimulates thyroid secretion of thyroxine** (T_4) which is converted in peripheral tissues (mainly liver and kidney) to more potent triiodothyronine (T_3)

    c.   Pathophysiology

        (1)   **Primary hypothyroidism — elevation of plasma TSH levels** with **low plasma thyroxine and triiodothyronine levels** secondary to primary thyroid disease; hyperplasia of pituitary thyrotrope cells occurs in long-standing cases

            (a)   Clinical symptoms — **decreased basal metabolic rate**, bradycardia, **sluggishness, constipation, weight gain, cold intolerance**, dry skin and hair, and **myxedema** (swelling due to abnormal mucin deposition in skin)

            (b)   **Cretinism** — hypothyroidism in infancy associated with **mental retardation**

        (2)   **Hypothalamic hypothyroidism** — damage to hypothalamus or pituitary stalk results in hypothyroidism (low plasma thyroxine and triiodothyronine levels), low or undetectable plasma TSH levels, and exaggerated TSH response to exogenous thyrotropin releasing hormone administration

        (3)   **Hyperthyroidism**

            (a)   TSH-secreting pituitary adenomas are rare; most cases of **thyrotoxicosis** are due to primary thyroid disease (**thyroiditis, toxic nodular goiter, Graves' disease**)

            (b)   Clinical symptoms — **tachycardia, tremor**, weight loss despite increased appetite, excessive sweating, and **heat intolerance**

3. **Gonadotropins**

  a. **Glycoprotein** hormones follicle-stimulating hormone (FSH) and luteinizing hormone (LH) composed of α-subunit (identical in both hormones) and ß-subunit (unique for each hormone); human chorionic gonadotrophin (HCG) secreted by placental syncytiotrophoblasts has identical α-subunit and unique ß-subunit

  b. **Stimulates gonads to form germ cells and secrete sex steroid hormones** (estradiol, progesterone, or testosterone)

    (1) **FSH** acts on **ovarian follicular cells** or **testicular Sertoli cells**

    (2) **LH** acts on **ovarian corpus luteum cells** or **testicular Leydig cells**

  c. Gonadotropins are secreted at low levels during childhood, but at puberty hourly secretory burst pattern begins

    (1) Males have relatively uniform basal "tonic" secretory burst activity

    (2) Females have "ovulatory surge" superimposed on basal tonic secretory activity

    (3) Postmenopausal females have high serum levels of gonadotropins

  d. Pathophysiology

    (1) **Hypergonadotropic hypogonadism — primary gonadal disease** (including castration) with consequent failure of sex steroid hormone secretion and elevation of pituitary gonadotropin secretion

    (2) **Hypogonadotropic hypogonadism — low gonadotropic secretion** with consequent failure of sex steroid hormone production; can occur from damage to hypothalamus or pituitary stalk or from various cerebral disorders or psychogenic conditions

      (a) **Delayed puberty** — often associated with obesity (adiposogenital dystrophy)

(b) **Secondary amenorrhea** — cessation of menstrual cycles after establishment of normal cycles in nonpregnant woman; causes other than structural disease, include depression, hypothyroidism, and medications (particularly psychotropic drugs such as phenothiazines)

(c) **Kallmann's syndrome** — **gonadal hypoplasia** and **anosmia** (loss of smell) associated with olfactory tract maldevelopment (arrhinencephaly) and color blindness; **X-linked inheritance** with incomplete penetrance: males have full syndrome while females have normal sexual function

(d) **Precocious puberty** — associated with pineal region tumors, hypothalamic hamartoma, or hypothalamic region tumors

4. **Human growth hormone (hGH)**

a. **Polypeptide hormone** acting at multiple sites and with numerous properties including **stimulation of musculoskeletal growth**, promotion of cellular amino acid uptake, and regulation of lipolysis and carbohydrate and mineral metabolism

(1) Produces hyperglycemia by **antagonizing effects of insulin (release of hGH is stimulated by hypoglycemia)**

(2) Promotes release of free fatty acids into circulation and increased hepatic fatty acid oxidation

b. **Secreted in bursts, particularly during early hours of sleep**

c. Short plasma half-life with most **effects being mediated by liver-synthesized somatomedin C (insulin-like growth factor-I; IGF-I)**

d. Pathophysiology

(1) **Growth hormone deficiency**

(a) Failure to increase plasma hGH levels in response to hypoglycemia or to administration of arginine or levodopa (L-Dopa)

(b)   Adults — no recognizable clinical symptoms

(c)   Children - **growth failure or dwarfism** and (in severe deficiency) **hypoglycemia** (due to loss of anti-insulin effect)

      i)   Growth hormone deficient dwarfs — respond to exogenous hGH administration

      ii)   **Psychosocial dwarfism** (maternal deprivation syndrome) — impaired hGH secretion (due to emotional factors) that can be rapidly returned to normal by placing child in supportive environment

      iii)   **Laron dwarfism — elevated hGH levels** and **low somatomedin C levels**, presumably due to defective tissue hGH receptors

(2)   **Excess growth hormone secretion** — plasma hGH levels above 5 ng/mL with failure to suppress secretion following glucose administration

(a)   **Gigantism** — prior to closure of bony epiphyseal plates, excess hGH secretion results in accelerated bone growth, particularly in long bones

(b)   **Acromegaly** — after bony epiphyseal closure, excess hGH secretion results in bone growth involving skull, face, and distal extremities producing characteristic appearance of **protruding jaw**, enlarged nose, **bulging forehead**, and **enlarged hands and feet**, along with enlarged viscera, skin thickening, and diabetes mellitus

5.   **Prolactin**

a.   **Polypeptide hormone** that **initiates and maintains milk production** by breast tissue primed by female sex steroid hormones

b.   **Hyperprolactinemia** — characterized by **galactorrhea** (breast discharge) and **amenorrhea** in females, or **gynecomastia** (breast enlargement), **galactorrhea**, and **impotence** in males

(1) **Pituitary isolation** — lesions of pituitary stalk or hypothalamus impair dopaminergic inhibition of prolactin secretion, resulting in increased plasma prolactin levels

(2) **Drugs** — interference with dopaminergic transmission by medications (such as chlorpromazine, haloperidol, or metoclopramide) results in increased plasma prolactin levels

(3) **Prolactin-secreting adenoma** — high serum prolactin levels (greater than 150 ng/mL) usually indicate prolactin-secreting adenoma

II. Pathology

A. **Sheehan's syndrome — hemorrhagic infarction of hyperplastic anterior pituitary gland of pregnancy** due to **intrapartum hypotension or shock** resulting in postpartum pituitary insufficiency

B. **Lymphocytic hypophysitis — autoimmune disorder** characterized by lymphoplasmacytic infiltration of anterior pituitary gland occurring mainly in women (usually during pregnancy) and can be associated with similar involvement of other endocrine organs; results in pituitary insufficiency

C. **Granulomatous hypophysitis** — non-caseating granulomas of anterior pituitary gland resulting in parenchymal destruction and pituitary insufficiency; must be differentiated from tuberculosis, fungal infection, or sarcoidosis

D. **Rathke's cleft cyst** — derived from Rathke's cleft; contains mucinous fluid and is lined by ciliated and goblet epithelial cells; common incidental finding at autopsy, but can be symptomatic when large enough to compress pituitary stalk or optic chiasm

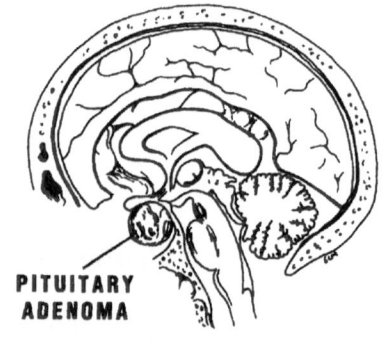

E. **Empty sella syndrome — developmental defect in diaphragma sellae** allowing arachnoid to descend into sella and push pituitary gland aside; usually asymptomatic radiologic curiosity, but increased cerebrospinal fluid pressure can produce traction on pituitary stalk with resultant hyperprolactinemia and reduction in secretion of other hormones

Pituitary adenoma enlarging sella turcica and elevating optic chiasm.

F. **Pituitary adenoma**

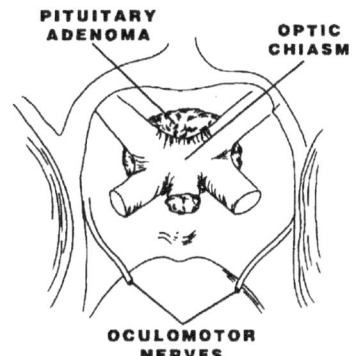

Superior view showing pituitary adenoma penetrating diaphragma sellae to elevate and compress optic chiasm.

1. Benign tumor of pituitary gland

    a. Classification based upon radiologic imaging size into microadenoma (less than 10 mm in diameter) or macroadenoma

    b. Endocrinologic classification into non-functioning (null cell) or hypersecreting (producing pituitary hormones)

2. Clinical presentations

    a. Non-functioning tumors

        (1) Tend to be larger (usually macroadenomas), since they lack endocrinologic symptoms and only become symptomatic when mass effect damages surrounding tissues

        (2) Visual disturbance — impingement on optic chiasm from below results in **bitemporal hemianopsia** (initially, bitemporal upper quadrants are more affected)

    b. Functioning tumors present with endocrine disturbances

3. Pathologic features

    a. Resembles normal pituitary gland except that each cell no longer contacts capillary for hormone secretion; identifiable in reticulin stains as expansion of normal reticulin network

    b. Some non-functioning tumors, that show histologic and biochemical evidence of hormone production, apparently have defective secretory ability

4. Most common tumor types are:

    a. **Prolactin cell adenoma**

    b. **Plurihormonal cell adenoma** — producing hGH, prolactin, and glycoprotein hormones

  c. **Corticotroph adenoma**

  d. **Null cell adenoma**

    (1) No hormone production

    (2) **Oncocytic** — subgroup of null cell adenoma in which cells contain abundant mitochondria

5. **Pituitary apoplexy** — acute severe headache, ophthalmoplegia, loss of vision, subarachnoid hemorrhage, and lethargy or coma due to **hemorrhagic infarction** of pituitary adenoma; can be life threatening and requires immediate treatment with high-dose intravenous corticosteroids and surgical decompression

6. Treatment

  a. Surgical resection (transcranial or transsphenoidal) followed by radiation therapy if residual tumor remains

  b. Postoperative hormone replacement is necessary if hormonal deficiency develops

  c. Since **dopamine** (released by hypothalamus) is normal inhibitor of prolactin release from pituitary cells, bromocriptine (dopamine receptor agonist) has been used to reduce prolactin levels and shrink prolactin-secreting tumors

SUGGESTED ADDITIONAL READING

Hankinson J, Banna M: *Pituitary and Parapituitary Tumours.* London, W. B. Saunders Co., 1976.

Kovacs K, Horvath E: *Tumors of the Pituitary Gland. Atlas of Tumor Pathology. Second Series. Fascicle 21.* Washington, DC, Armed Forces Institute of Pathology, 1986.

Lloyd RV (ed): *Surgical Pathology of the Pituitary Gland.* Philadelphia, W. B. Saunders Co., 1993.

Martin JB, Reichlin S: *Clinical Neuroendocrinology.* Edition 2. Philadelphia, F. A. Davis Co., 1987.

CHAPTER 17: EYE PATHOLOGY

I.   External structures

A.   **Orbit** — contains a variety of tissues including skeletal muscle (rectus and oblique muscles), lacrimal glands, optic nerve, peripheral nerves, blood vessels, lymphoid tissue, and fibroadipose tissue; pathologic processes include:

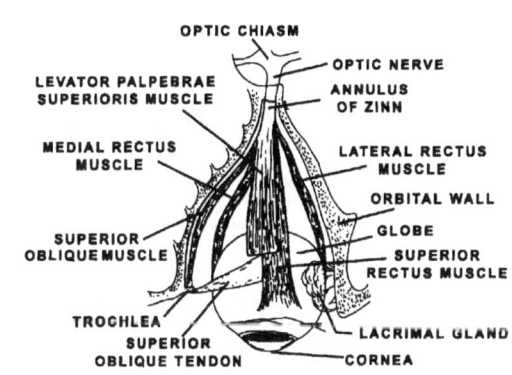

Left orbit from above.

1.   **Neoplasm** — Primary tumors can arise from any orbital tissue or can be involved secondarily by tumor invasion from adjacent structures or from distant metastases

2.   **Inflammation**

a.   Acute inflammation — results from orbital injuries or by extension from adjacent structures (globe, sinuses, teeth, middle ear, or cranial cavity)

b.   **Chronic inflammation (orbital pseudotumor)** — usually of unknown etiology, but rare specific etiologies include tuberculosis, fungi, and parasites

3.   **Endocrine exophthalmos**

a.   Increase in orbital tissue mass due to **edema, muco-polysaccharide accumulation, lymphocytic inflammation,** or **fibrosis**

b. Associated with thyroid disorders (**"thyroid eye disease"**)

c. Earliest clinical manifestation being slight lid retraction exposing sclera superiorly

d. **Malignant exophthalmos** — progressive lid edema, conjunctival chemosis, and severe exophthalmos with exposure keratitis, immobility of globe (with rubbery, pale, hypertrophic extraocular muscles), visual loss, glaucoma, and panophthalmitis

B. **Eyelids**

1. Composed of four layers: external cutaneous layer, subjacent muscular layer, deeper **tarsal plate** (dense collagenous and elastic tissue), and deepest inner lining of **palpebral conjunctiva**

2. **Accessory glands** — sebaceous (Zeis) and serous (Moll) glands empty into follicles of eyelashes

3. Meibomian glands — special sebaceous glands lying in tarsal plate and emptying at mucocutaneous junction

4. Tears washing conjunctiva drain through lacrimal canaliculi, into lacrimal sac and then through nasolacrimal duct into nasal passages

5. Pathologic changes are those of skin or conjunctiva:

a. **Hordeolum (stye)** — localized **infection** (frequently staphylococcal) of accessory glands or Meibomian glands

b. **Chalazion** — non-infectious granulomatous inflammation surrounding lipid droplets (**lipogranuloma**) in sebaceous (Meibomian or Zeis) glands

c. **Dacryocystitis** — suppurative inflammation of lacrimal sac superimposed on obstruction to tear flow

d. **Entropion** — **inversion** of eyelid margin; results in inflammation of conjunctiva (conjunctivitis) and cornea (keratitis) from scraping by inverted lashes

e. **Ectropion** — **eversion** of eyelid margin; results in conjunctivitis and keratitis from drying of exposed conjunctiva and cornea

    f.   **Xanthelasma** — focal (often perivascular) collection of lipid-filled macrophages in dermis of eyelid skin; related to systemic lipid abnormalities (hypercholesterolemia)

    g.   **Basal cell carcinoma** or **squamous cell carcinoma** of eyelid skin

II.   Globe

  A.   Landmarks

      1.   **Anterior pole** — center of cornea

      2.   **Posterior pole** — point of greatest curvature of posterior segment; optic nerve is 3 to 4 mm nasally from posterior pole and 1 mm below horizontal meridian

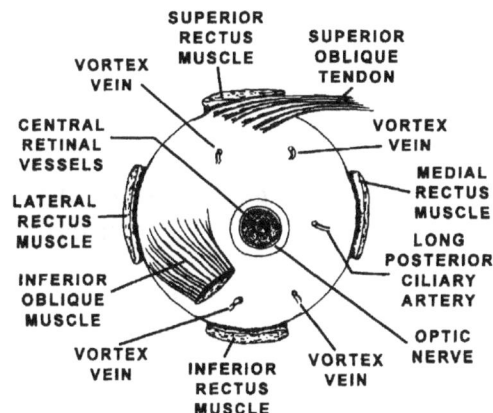

Posterior view of right globe.

      3.   Anatomical axis — largest diameter (24 mm) between anterior and posterior poles

      4.   Optical axis — straight line between centers of curvatures of refractive surfaces (cornea and lens)

      5.   **Visual axis** — straight line between fovea (macula) and fixation point

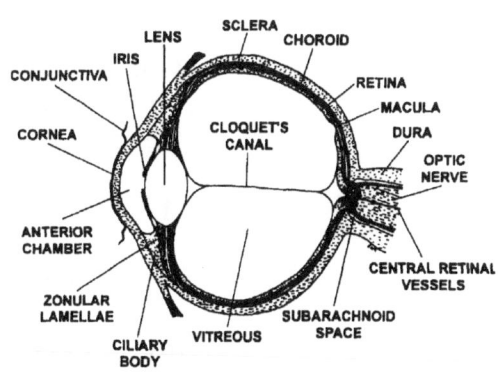

  B.   **Cornea, sclera,** and **conjunctiva**

      1.   Cornea forms anterior one-fifth and sclera forms posterior four-fifths of outer fibrous envelope of eye

Horizontal section of globe.

      2.   **Limbus** — transition zone between transparent cornea and opaque sclera

3. **Cornea**

    a.   Optical properties depend on surface smoothness, curvature, and transparency; normally there are no corneal blood vessels, but numerous corneal sensory nerve endings

    b.   **Corneal epithelium**

        (1)   Anterior corneal surface of **non-keratinizing stratified squamous epithelium** covered by surface tear film

        (2)   Serves as major refracting ocular surface, accounting for 46 out of total 50 diopters of ocular refractive power

    c.   **Corneal stroma** — 90% of total corneal thickness, consisting of **regularly arrayed collagen fibrils** and fibroblasts **(keratocytes)** embedded in mucopolysaccharide ground substance; transparency is maintained by careful **regulation of water content**

        (1)   **Bowman's membrane — anterior** acellular layer of collagen fibrils, located just **beneath corneal epithelium**

        (2)   **Descemet's membrane** — thick **posterior** acellular collagenous layer, located at **base of corneal endothelium**

    d.   **Corneal endothelium** — posterior corneal surface formed by single layer of flat, hexagonal cells whose apical surface is bathed by anterior chamber aqueous humor

        (1)   To assure stromal transparency corneal endothelium regulates water transport, maintaining relatively reduced hydration

        (2)   Loss of corneal endothelium (as occurs with aging or following surgery on globe) results in loss of normal stromal water regulation with consequent corneal clouding

4. **Sclera**

    a.   Opaque fibrous envelope comprising posterior outer surface of globe

b.   Composed of **randomly-arranged, wide bundles of collagen fibrils** that taper and split into intimately interwoven bands producing felt-like structure with great tensile strength, extensibility, and flexibility in all directions enabling it to equalize variations in intraocular pressure

c.   **Episclera** — superficial scleral stroma that has looser, finer collagen fibrils and connects to conjunctiva.

d.   Optic nerve dural sheath is continuous with sclera

e.   **Tenon's capsule** (fascia bulbi) — thin, fibrous membrane of delicate collagenous trabeculae connecting episclera to orbital muscle sheaths, retrobulbar fat, and conjunctiva

5.   **Conjunctiva**

a.   **Mucous membrane** covering posterior surface of eyelids (**palpebral conjunctiva**) and anterior surface of globe (**bulbar conjunctiva**)

b.   Contains rich vascular supply and lymphatic drainage ending at limbus

c.   Composed of **stratified columnar epithelium** (containing **goblet cells** that produce mucoid layer of tear film) which is continuous with stratified squamous corneal epithelium

6.   Pathologic processes

a.   **Epithelial neoplasms** (**squamous carcinoma** or **adenocarcinoma**) can arise from conjunctival or corneal epithelium

b.   **Corneal dystrophy** — degenerative corneal disorders characterized by structural distortion resulting in loss of transparency (corneal clouding) and consequent diminished vision

(1)   **Keratomalacia** — epithelial keratinization due to **vitamin A deficiency**

(2)   **Keratoconjunctivitis sicca** — epithelial damage from deficient tearing; part of systemic disorder (**Sjögren's**

syndrome) characterized by dry eyes, mouth, nose, and vagina, achlorhydria, and chronic polyarthritis

(3) **Neurotrophic keratopathy — damage to ophthalmic branch of trigeminal nerve** (V_1) results in loss of corneal sensation and loss of trophic influences on corneal epithelial cells which no longer resist minor trauma, desiccation, or infection; consequent edema, vesicle formation, and exfoliation produce corneal scarring; sewing eyelids together prevents corneal damage

(4) **Band keratopathy — calcium deposition in Bowman's membrane** with underlying fibrosis of corneal stroma producing horizontal cloudy band that progresses from limbus on each side to center of cornea; associated with **hypercalcemia**

(5) **Hereditary corneal dystrophy** — group of heredofamilial disorders resulting in corneal clouding of both eyes but without any identifiable systemic disease

(6) **Keratoconus** — fragmentation and thinning of Bowman's membrane and adjacent corneal stroma in central cornea resulting in cone-like outward corneal protrusion; mechanical damage to corneal endothelium alters stromal hydration with consequent corneal cloudiness

c. **Kayser-Fleischer ring** — bright green or red circumferential corneal band near limbus resulting from **copper** sulfide deposition in **Descemet's membrane**; correlates with **neurologic symptoms** in **Wilson's disease (hepatolenticular degeneration)**

d. **Pinguecula and pterygium**

(1) **Yellow elevated area of conjunctiva** close to limbus (**pinguecula**) and wing-shaped mass of **thickened conjunctiva progressively extending across limbus onto cornea (pterygium)**

(2) Characterized by epithelial irregularity, degeneration of basement membrane (Bowman's membrane over corneal portion of lesion), hyalinized degenerating collagen fibers, fibroblastic proliferation, and increased vascularity

(3)   Related to **irritation** from sunlight (ultraviolet irradiation), particulate matter (dust), wind, or chemicals

e.   **Inflammation**

(1)   Rheumatoid scleritis — granulomatous inflammation of sclera resulting in abnormal thickening (brawny scleritis) or thinning and perforation (scleromalacia)

(2)   Corneal inflammation (**keratitis**) is nearly always accompanied by **conjunctivitis** (keratoconjunctivitis), but primary conjunctival inflammation (conjunctivitis) may be accompanied by little or no keratitis

(3)   Keratitis, conjunctivitis, or keratoconjunctivitis range from mild hyperemia and edema to severe suppuration; etiologic agents are numerous and varied:

   (a)   Physical agents including simple drying, particulate matter, trauma, or ionizing radiation

   (b)   Chemical agents including toxic, irritating, or caustic substances

   (c)   Allergic reactions such as those to pollens in hay fever or to drugs and other chemicals

   (d)   Bacterial agents

      i)   Many kinds of bacteria produce conjunctivitis (e.g. pneumococcus, *Hemophilus influenzae*, Streptococcus, and Staphylococcus)

      ii)   **Gonococcus** (*Neisseria gonorrhea*) produces severe **suppurative keratoconjunctivitis in infants** infected during birth process, with resultant blindness unless promptly treated

   (e)   **Trachoma** (chlamydia)

      i)   Viral agent that initially involves conjunctiva but extends to cornea with ultimate scarring and blindness

ii) Rare in North America, but most common cause of blindness worldwide

(f) **Herpes simplex keratitis**

    i) Superficial (dendritic) ulcers characterized by intracellular edema, necrosis, and ulceration of epithelium similar to that of dermal lesions

    ii) Deep destructive lesions (disciform keratitis) characterized by stromal edema and dissolution of collagen fibers with potential perforation and scarring

(g) Interstitial keratitis — associated with congenital syphilis

(4) **Corneal ulcers** — corneal epithelial injury with exposure of corneal stroma; sequellae include:

(a) Distortion and scarring of cornea

(b) **Hypopyon** — inflammation in anterior chamber

(c) Secondary glaucoma — obstruction of aqueous outflow by inflammatory products

(d) Corneal perforation with collapse of anterior chamber

(e) **Endophthalmitis** or **panophthalmitis** — extension of infection with ultimate disruption of internal ocular structures and atrophy

(f) **Phthisis** — disruption and destruction of internal ocular structures which are replaced by fibrosis

C. Anterior and posterior chambers

1. **Anterior chamber**

    a. Compartment bounded anteriorly by cornea, peripherally by trabecular meshwork and angle tissues, and posteriorly by iris and lens

b. Communicates with posterior chamber through pupil

2. **Posterior chamber**

 a. Roughly triangular compartment limited anteriorly by posterior iris surfaces, ciliary body laterally, and anterior vitreous posteriorly

 b. Lens occupies portion of posterior chamber and is separated from vitreous by canal of Petit filled with aqueous humor

3. Anterior chamber angle

 a. Angle ("filtration angle") between cornea, ciliary body, and iris, consisting of honeycomb series of small trabeculae (trabecular meshwork) forming exit channels to drain aqueous humor from anterior chamber into scleral venous system (Schlemm's canal)

 b. Scleral spur — anterior termination of sclera adjacent to anterior chamber angle

4. **Intraocular pressure** — dependent upon balance between formation of aqueous humor by ciliary body (in posterior chamber) and its drainage through trabecular meshwork; normal production rate is 2 μL/min and normal intraocular pressure is less than 30 torr (mm Hg)

 a. **Decreased intraocular pressure** — results from impaired aqueous formation (by ciliary body); associated with subsequent atrophy or phthisis

 b. **Glaucoma (increased intraocular pressure)** — results from impaired aqueous flow (around lens, through pupil, or through trabecular meshwork into Schlemm's canal); pathologic changes secondary to increased intraocular pressure include retinal, uveal, and scleral atrophy, and cupping of optic nerve head

  (1) **Open-angle glaucoma — insidious onset** with no symptoms until visual impairment occurs; results from degenerative alterations in trabecular meshwork with progressive outflow obstruction; genetic predisposition necessitates testing of relatives

  (2) **Angle-closure (narrow angle) glaucoma** — congenital narrow angle predisposes to **sudden onset** of blurred

vision and pain following any process that thickens base of iris or causes iris to press forward against trabecular meshwork (such as iris edema or sudden pupillary dilation); adhesions between peripheral iris and posterior corneal surface permanently obstruct aqueous outflow

(3) **Secondary glaucoma** — secondary to some other ocular disease such as inflammation or neoplasm

(4) **Congenital glaucoma** — increased intraocular pressure in infancy results in **enlargement of globe (buphthalmos)** due to stretching of all layers

5. **Inflammation** — extensive intraocular inflammation is most commonly secondary to penetrating wounds

   a. **Hypopyon** — purulent inflammatory debris in anterior chamber; may be associated with blood in anterior chamber (hyphema)

   b. **Endophthalmitis** — inflammatory process involving anterior and posterior chambers, vitreous, uvea, and retina, but sparing sclera and orbit

   c. **Panophthalmitis** — inflammation involving sclera and orbital tissues in addition to anterior and posterior chambers, vitreous, uvea, and retina

   d. **Synechia** — intraocular scarring (**fibrous adhesion**) between iris and cornea (iridocorneal) or lens and iris (iridolenticular)

   e. **Cyclitic membrane** — membrane-like scar formation attaching vitreous body circumferentially to ciliary body resulting in reduced aqueous production and decreased intraocular pressure with eventual atrophy of globe

   f. **Phthisis bulbi** — following formation of cyclitic membrane with fibrotic attachments to vitreous and retina, scar contraction results in detachment of retina, ciliary body, and choroid and shrinkage of globe; disorganized internal ocular structures characteristically undergo **osseous metaplasia (bone formation)**

D. **Uveal tract** — choroid, ciliary body, and iris

1. **Choroid**

    a. Posterior portion of uveal tract lying external to retinal pigmentary epithelium and internal to sclera consisting of blood vessels, nerve fibers, and pigmented cells (**melanocytes**) set in loose connective tissue matrix

    b. Nourishes retina and shields retina from extraneous light

        (1) Supplied by branches of posterior and anterior ciliary vessels (derived from ophthalmic artery) and drained by vortex and ophthalmic veins

        (2) Three layers of choroidal vessels:

            (a) Largest and outermost vessels (Haller's layer)

            (b) Intermediate-sized vessels (Sattler's layer) internal to Haller's layer

            (c) Innermost **choriocapillaris** (choriocapillary network) situated adjacent to retinal pigmentary epithelium

        (3) **Bruch's membrane** — basal lamina of both choriocapillaris and retinal pigmentary epithelium, which acts as **diffusion barrier** to passage of molecules from choriocapillaris into retina

2. **Ciliary body**

    a. Extends from ora serrata (periphery of retina) to iris root

    b. Contains ciliary muscles, nerves, and vasculature, along with **neuroectodermal** pigmented (external) ciliary epithelium (continuation of retinal pigmentary epithelium), and **neuroectodermal** (internal) nonpigmented ciliary epithelium (continuation of neurosensory retina)

    c. Produces aqueous humor

    d. Has basal lamina containing zonular fibers connected to lens

3. **Iris**

    a. Anterior part of uvea

    b. **Anterior layer** — contains fibroblasts, melanocytes, and nerve cells; iris color depends on thickness and amount of pigment in this layer

    c. **Pigmented anterior (external) iris epithelium** — continuation of pigmented ciliary epithelium; gives rise to iris muscles

        (1) **Sphincter (constrictor) muscle** — smooth muscle strip encircling pupillary margin producing **pupillary constriction (parasympathetic** innervation)

        (2) **Dilator muscle** — continuous smooth muscle sheath producing **pupillary dilation (sympathetic** innervation)

    d. Pigmented posterior (internal) iris epithelium — continuation of nonpigmented ciliary epithelium

4. Pathologic processes

    a. **Inflammation**

        (1) **Anterior uveitis**

            (a) Involving iris (iritis), ciliary body (cyclitis), or both (iridocyclitis)

            (b) Causes include idiopathic, viral (cytomegalovirus, herpes simplex, varicella-zoster, measles), bacterial (tuberculosis, syphilis, leprosy, brucellosis, leptospirosis), fungal (histoplasmosis, actinomycosis, blastomycosis), and systemic diseases (rheumatoid arthritis, Behçet's disease, Boeck's sarcoid, Hodgkin's disease)

        (2) **Posterior uveitis**

            (a) Involving choroid (choroiditis) and often also involving retina (**chorioretinitis**)

    (b)   Causes include idiopathic, toxoplasmosis, histoplasmosis, and collagen-vascular diseases, and systemic diseases

(3)   **Panuveitis**

    (a)   Combined anterior and posterior uveitis

    (b)   **Sympathetic ophthalmitis** — immune-mediated granulomatous reaction throughout uveal tract (following perforating injury to eye) that may also involve uninjured eye, usually after several weeks, but even after many years

(4)   Rubeosis iridis — neovascularization (fibrovascular tissue) on the anterior surface of iris

b.   **Melanoma**

(1)   Most common uveal tumor (80% originate in **choroid**)

(2)   Varies from **low grade** malignancy (**spindle cell type**) to **high grade** malignancy (**epithelioid type**)

(3)   Prognosis depends on tumor size, presence or absence of extension through perivascular and perineural spaces in sclera, and presence or absence of blood borne metastases

E.   **Lens**

1.   Transparent biconvex structure that varies in shape with resultant change in refractive properties

a.   **Capsule** — encircling membrane that provides insertion for zonular fibers

b.   **Zonular fibers**

(1)   Originate in basal lamina of neuroectodermal nonpigmented ciliary epithelium and transmit ciliary muscle contraction to lens during accommodation (near focusing)

(2) **Ciliary muscle contraction** makes zonular **fibers lax** causing **lens to become thicker** with more refractive power

c. **Epithelium** — cells that produce lens fibers lying immediately below anterior lens capsule; anterior cell nuclei form single row immediately below capsule, but near lens equator (where new lens fibers are formed) nuclei are pushed inward resulting in "nuclear bow" shape

d. **Lens fibers** — long cytoplasmic processes of epithelial cells; growth of lens throughout life occurs from increase in number of fibers; as fibers are added at periphery (lens cortex), deeper fibers (lens nucleus) lose water to become firm and sclerotic, resulting in **loss of lens pliability** (clinically evident as **presbyopia**) and darkening of lens nucleus (nuclear cataract)

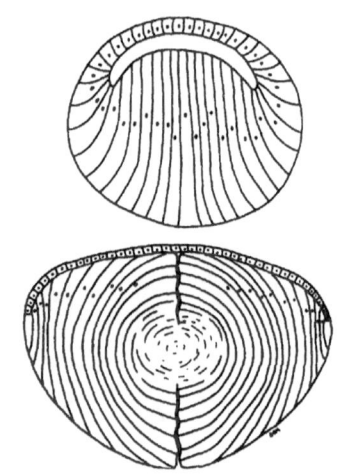

Developing lens (top) and mature lens (bottom) showing nuclear bow and sclerotic nucleus.

2. Pathologic process

a. **Cataract** — loss of transparency of lens

(1) **Nuclear cataract** — normal part of aging accompanied by deepening yellow to amber color

(2) **Cortical cataract** — damage to lens fibers with consequent water loss and shrinkage resulting in clefts, vacuoles, fragmentation, and loss of transparency; causes include metabolic alterations (especially those associated with diabetes and aging), radiant energy (both ionizing and non-ionizing), trauma, and inflammation in surrounding ocular structures

(a) **Morgagnian globules** — eosinophilic protein in vacuoles or clefts between degenerating lens fibers

(b) **Bladder cells** — swollen vesicular epithelial cells in equatorial region

(3) **Subcapsular cataract** — injury to anterior subcapsular epithelial cells (usually from adjacent inflammation) results in proliferation and piling up of cells and fibroblastic transformation with consequent **collagen production**

F. **Vitreous**

1. Relatively acellular transparent viscoelastic **gel** composed of **collagen fibrils** embedded in protein-hyaluronic acid matrix

2. Supports retina and mechanically dampens movements or sudden shocks

3. **Cloquet's canal** — central soft area of vitreous; remnant of embryonic space occupied by hyaloid blood vessels during early development

G. **Retina**

1. Two delicate layers (**sensory retina** and **retinal pigmentary epithelium**) derived embryologically from same outgrowth of optic cup and lying in close apposition to each other but not mechanically attached except at ora serrata (extreme periphery of retina) and at optic disc

2. Vascular supply

   a. Internal retinal layers are supplied by branches of **central retinal artery and vein**

   b. Outer retinal layers (photoreceptors and retinal pigmentary epithelium) are nourished by **choriocapillaris**

ERG: large amplitude, long latency rod response and short latency, low amplitude cone response.

3. Electrophysiology

   a. **Electroretinography** (ERG)

      (1) Electrical response (recorded by contact lens electrode) of rods and cones to light stimulation

(2) Rods are stimulated by low-intensity blue light in dark background and cones are stimulated with high intensity white light in background of steady white light

(3) Rods have larger amplitude slower response than cones

b. Visual evoked potentials (VEP) — scalp-recorded electrical signal generated in occipital cortex in response to visual stimulation

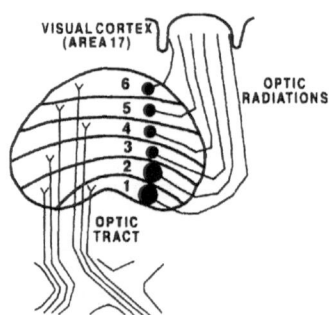

Lateral geniculate nucleus input and projections.

(1) Fibers from each eye join at optic chiasm to travel in optic tract to lateral geniculate, which then projects to visual (calcarine) cortex (area 17)

(a) Lateral geniculate layers 1, 4, 6 receive fibers from contralateral eye, while layers 2, 3, 5 receive fibers from ipsilateral eye

(b) Magnocellular layers (layers 1 and 2) mediate location and motion

(c) Parvicellular layers (layers 3, 4, 5, and 6) mediate color and form

(2) Flash visual evoked potential — visual stimulation with flickering strobe light; responses are too variable for clinical utility

(3) Pattern-reversal visual evoked potential — visual stimulation with checkerboard pattern of black and white squares that repeatedly reverse position

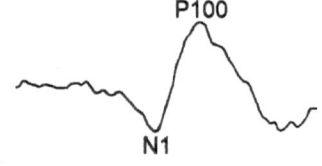

Pattern-reversal VEP with P100 positive peak at 100 milliseconds.

(a) Potential has initial negative peak (N1) followed by **large positive peak (P100)** that occurs with latency of approximately 100 milliseconds

    (b)   Difference in latency following stimulation of each eye is sensitive indicator of optic nerve pathology as occurs in optic neuritis or multiple sclerosis

4. Retinal pigmentary epithelium

a. Composed of pigmented cells containing **neuromelanin** (melanin in membrane-bound structures resembling secondary lysosomes; in contrast, melanin of choroidal melanocytes is contained in melanosomes)

Abnormal left VEP amplitude and latency indicates optic nerve pathology in patient with multiple sclerosis.

b. Apical surfaces of retinal pigmentary epithelial cells contain numerous microvilli that closely contact outer segments of photoreceptors (rods and cones) facilitating interchange of metabolites and visual pigment precursors

    (1)   Processes and recycles vitamin A that is bleached during normal vision

    (2)   Acts as barrier that regulates exchange of substances between choriocapillaris and photoreceptors

    (3)   Reduces random back scatter of light from sclera by absorption in melanin granules

    (4)   **Phagocytizes tips of photoreceptor outer segments,** which are constantly being shed and resynthesized

5. Fovea (macula)

a. 1.5 mm diameter area of retina (located 4 mm temporal to optic disc) containing greatest concentration of photoreceptors (cones only)

b. Retina is only half its usual thickness allowing photoreceptors direct exposure to incoming light without intermediary scattering that occurs elsewhere in retina

6.   **Optic disc** (optic nerve head) — rounded area where optic nerve pierces globe through scleral foramen; lacking all neuronal elements except ganglion cell axons, optic disc is physiologic blind spot in visual field testing

    a.   **Papilla** — retinal elevation from ganglion cell axons (approximately 1,200,000/eye) becoming crowded near optic disc before making 90 degree bend to enter optic nerve

    b.   **Physiologic cup** — central indentation of optic disc; filled with astrocytes that are remnants of glial tissues originally surrounding embryonic hyaloid artery

    c.   **Lamina cribrosa** — sieve-like network of collagenous fibers (extensions of sclera) penetrating optic nerve, running between groups of glia-covered axons

    d.   **Central retinal artery** — branch of ophthalmic artery that enters optic nerve behind globe from below; within optic nerve central retinal artery and vein run parallel to meninges and divide near disc surface to form superior and inferior retinal vascular arcades

    e.   **Optic nerve sheath** — extensions of dura and leptomeninges surround optic nerve and fuse with sclera; cerebrospinal fluid in subarachnoid space surrounding optic nerve freely communicates with that in subarachnoid space surrounding brain

7.   **Sensory retina** — transduces light into electrical signals; photoreceptor signals are integrated in retina and ultimately conveyed through optic nerve to brain

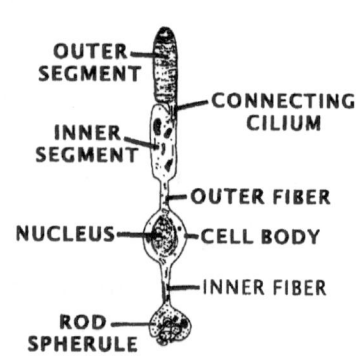

    a.   **Photoreceptors**

        (1)   **Rods** — approximately 120 million cells per eye dispersed throughout retina; contain rhodopsin visual pigment for resolving dim light in dark background (**scotopic** conditions)

Photoreceptor (rod).

(2) **Cones** — about 7 million cells concentrated mainly in macula and surrounding retina; three subpopulations containing visual pigments sensitive to red, green, or blue light; responsible for fine resolution and color perception in adequate (bright) background lighting (**photopic** conditions)

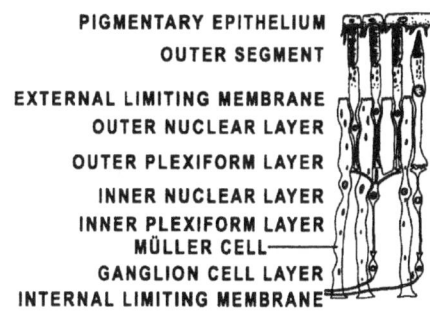

PIGMENTARY EPITHELIUM
OUTER SEGMENT
EXTERNAL LIMITING MEMBRANE
OUTER NUCLEAR LAYER
OUTER PLEXIFORM LAYER
INNER NUCLEAR LAYER
INNER PLEXIFORM LAYER
MÜLLER CELL
GANGLION CELL LAYER
INTERNAL LIMITING MEMBRANE

Relationship of retinal photoreceptors and neurons.

b. Retinal layers

(1) **External limiting membrane** — row of junctional complexes between photoreceptor cytoplasm and **Müller cell (glial) cytoplasm**

   (a) Space between external limiting membrane and retinal pigmentary epithelium contains photoreceptor inner and outer segments such that outer segment tips come into close contact with apical microvilli of retinal pigmentary epithelial cells

   (b) Each retinal pigmentary epithelial cell contacts 20-30 photoreceptor outer segments

(2) Outer nuclear layer — **photoreceptor nuclei**

(3) Plexiform layer — axons of photoreceptor and horizontal cells and dendrites of bipolar cells

(4) Inner nuclear layer — nuclei of retinal neuronal and glial cells

   (a) Horizontal cells — complex arborizing processes that synapse with photoreceptor axons and bipolar cell dendrites

        (b)   **Bipolar cells** — dendrites contact photoreceptors and axons synapse with ganglion cell dendrites

        (c)   Amacrine cells — pear-shaped cells that have arborizing processes synapsing with ganglion cells

        (d)   **Müller cells** — retinal glial cells that provide structural support with fibers extending to form both internal and external limiting membranes

    (5)   Inner plexiform layer — axons of bipolar and amacrine cells and dendrites of ganglion cells

    (6)   **Ganglion cell layer — cell bodies of ganglion cells**

    (7)   **Nerve fiber layer — axons of ganglion cells** coursing toward optic disc to form optic nerve

    (8)   Internal limiting membrane — junctional complexes between Müller cell processes serving as boundary between retina and vitreous

 c.   **Light must pass through retina before it activates rods and cones**

    (1)   Any retinal alteration that reflects, refracts, or obstructs light passage reduces availability of light to photoreceptors

    (2)   Retinal pigmentary epithelium absorbs any light not absorbed by photoreceptors and hence diminishes scatter

8.   Pathologic process

 a.   Vascular disease

    (1)   Circular exudates and hemorrhages — since deeper (outer) retinal neural structures are oriented perpendicular to retinal surface, exudates and hemorrhages in deeper (outer) layers are usually seen as roughly circular, well-circumscribed foci

    (2)   **Flame-shaped exudates and hemorrhages** — since nerve fiber layer is oriented parallel to retinal surface and fibers converge toward optic disc, exudates and

hemorrhages in nerve fiber layer are elongated in direction of fibers producing flame shape

(3) **Cotton-wool exudates** — microinfarcts of nerve fiber layer

(4) **Diabetic retinopathy**

(a) Retinal artery arteriosclerosis can produce vascular occlusion and retinal infarcts

(b) Rupture of **capillary aneurysms** results in retinal hemorrhage; both aneurysms and hemorrhages appear by ophthalmoscopy as scattered small red retinal dots (hemorrhages being somewhat larger); resolution of hemorrhages results in masses of proteinaceous fluid and lipid visible as punctate, slightly yellow foci

(c) **Neovascularization** (formation of new capillaries) along inner surface of retina predisposes to hemorrhage into vitreous; organization of such hemorrhages results in preretinal scars which incite more new vessel formation (**retinitis proliferans**) predisposing to further hemorrhage; scar retraction can cause retinal detachment

(5) **Hypertensive retinopathy**

(a) Spasm and narrowing of arterioles in hypertension results in microinfarction accompanied by increased permeability of capillaries which in turn results in exudates and hemorrhages

(b) Small edematous **retinal infarcts (cotton wool exudates)**; these contain **cytoid bodies** which are swollen axons in nerve fiber layer

(c) **Retinal edema** commonly progresses to **cystoid degeneration** in deeper layers, and to deposition of eosinophilic masses appearing as round, white, sharply defined foci by ophthalmoscopy; focal separation (detachment) of sensory retina from retinal pigmentary epithelium can occur as fluid accumulates

b. **Retinal detachment** — since sensory retina is only attached to retinal pigmentary epithelium at optic disc and ciliary body, detachment or separation of sensory retina from retinal pigmentary epithelium results in conical mass with circumferential base at ciliary body and apex at optic disc

   (1) Separation of rods and cones from normal intimate relationship with retinal pigmentary epithelium and nutrients supplied by choriocapillaris results in photoreceptor degeneration

   (2) Mechanisms

      (a) **Accumulation of subretinal fluid** from inflammation in retina or choroid, vascular lesions in either choroid or retina, or extravasation from tumors

      (b) **Retraction of adherent fibrotic vitreous** secondary to organization of endophthalmitis or hemorrhages in space between vitreous and retina (preretinal or subhyaloid space)

      (c) **Holes or tears in retina** which allow vitreous fluid to escape into subretinal space; aging, myopia, retinal atrophy, cystoid degeneration, and trauma predispose to such tears

c. **Papilledema — optic disc swelling** secondary to **increased intracranial pressure**

   (1) Associated with absence of **spontaneous venous pulsations** in veins entering optic disc and small linear hemorrhages in nerve fiber layer near optic disc

   (2) Results from transmission of increased intracranial pressure by cerebrospinal fluid in subarachnoid space surrounding optic nerve

      (a) Distortion of lamina cribrosa from increased pressure **interferes with normal axoplasmic**

transport in optic nerve fibers resulting in axonal swelling which reduces size of physiologic optic cup and elevates optic disc margins

(b) Swollen nerve head protrudes forward against vitreous

(c) Lateral protrusion of swollen nerve head causes buckling of peripapillary retina with focal **retinal detachment** and accumulation of proteinaceous fluid between sensory retina and retinal pigmentary epithelium; these alterations result in **enlargement of blind spot** in visual field testing

(d) **Chronic papilledema** results in **axonal degeneration** and **glial proliferation**

d. **Pigmentary retinopathy** — associated with various disorders including retinitis pigmentosa, Kearns-Sayre syndrome (mitochondrial disorder), Refsum's disease, or ceroid lipofuscinosis

(1) Degeneration of photoreceptors stimulates degeneration and proliferation of retinal pigmentary epithelium; macrophages containing pigment phagocytized from degenerated retinal pigmentary epithelial cells migrate toward retinal veins, while proliferating retinal pigmentary epithelial cells migrate into superficial layers of sensory retina

(2) Loss of peripheral photoreceptors (rods) results in **night blindness**

(3) Electroretinography useful in diagnosis

(4) **Retinitis pigmentosa**

(a) **Hereditary pigmentary retinopathy** beginning in mid-retina (resulting in characteristic ring scotoma) and progressing peripherally and centrally

(b) **Bone spicule appearance** of pigment deposition by ophthalmoscopy

e.   **Retinoblastoma** (primitive neuroectodermal tumor of retina)

(1)   Common tumor of childhood presenting as **strabismus** (eye deviation), **leukokoria** ("cat's eye reflex" — **white pupillary reflex** due to mass in pupillary area behind lens)

(2)   Often **multifocal** within one eye or **bilateral**; when bilateral and in association with histologically-similar pineal tumor (**pineoblastoma**) has been termed "**trilateral retinoblastoma**"

(3)   **Familial** in 5% of cases; prototype for cancers caused by mutational loss of genetic information

(a)   "**Two-hit**" **model of tumorigenesis** — retinoblastoma arises due to two mutational events involving region on long arm of **chromosome number 13** (region **13q14**)

i)   **No tumor** develops if both chromosomes of pair have normal region 13q14 or if only one chromosome of pair has deletion of 13q14 region (**heterozygous for deletion**)

ii)   **Tumor develops** if both of chromosome 13 pair has deletion of 13q14 region (**homozygous for deletion**)

(b)   Hereditary (**autosomal dominant**) form — abnormal chromosome 13 with deleted 13q14 region is inherited from one parent; subsequent, mutation (with 90% probability) in 13q14 region of other chromosome 13 in some cells during development of eye, results in tumor growth from these abnormal cell clones

(c)   Sporadic form — mutational events resulting in deletion of 13q14 region in both of chromosome 13 pair in same cell occur at sometime during eye development resulting in tumor cell clone with resultant tumor growth

(4)   Treatment involves **enucleation**; cure is possible if tumor has not spread into optic nerve; familial cases have predisposition to **second malignancies** (particularly **osteosarcoma** of femur)

SUGGESTED ADDITIONAL READING

Hogan MJ, Zimmerman LE (eds): *Ophthalmic Pathology. An Atlas and Textbook*. Second Edition. Philadelphia, W. B. Saunders, 1962.

Farris BK (ed): *The Basics of Neuro-ophthalmology*. St. Louis, Mosby-Year Book, 1991.

Yanoff M, Fine BS: *Ocular Pathology: A Text and Atlas*. Third Edition. Philadelphia, J. B. Lippincott, 1989.

CHAPTER 18: EAR PATHOLOGY

I.   External ear

A.   **Pinna — elastic cartilage** covered with tightly apposed skin and minimal subcutaneous tissue (except for ear lobe which has fat pad between layers of skin)

B.   **External auditory canal**

1.   Superficial (external) portion

a.   Cartilage continuous with pinna

b.   Lining skin contains hair with sebaceous glands and **ceruminous (apocrine) glands** (which resemble axillary and pubic apocrine glands)

2.   Deep (internal) portion

a.   Bony structure

b.   Lining skin has no adnexal structures and has thin subcutaneous tissue that merges with periosteum

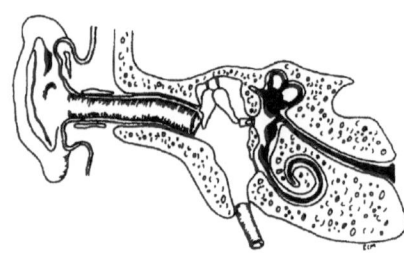

External, middle, and inner ear.

C.   **Tympanic membrane**

1.   Membrane separating external ear from middle ear

2.   External surface

a.   **Epidermal layer of stratified keratinizing squamous epithelium** continuous with skin of external auditory canal

      b.   **Auditory epithelial migration**

           (1)   Continuous movement of epithelium (and surface keratinized layer) over tympanic membrane and deep portion of external auditory canal toward surface of external (cartilagenous) portion of external auditory canal prevents accumulation of squamous debris

           (2)   **Keratoma (cholesteatoma) — growth of epithelium that has migrated into middle ear through perforation in tympanic membrane**

   3.   Central layer — **connective tissue** with external radially-oriented collagenous bundles and internal circularly arranged collagen bundles

   4.   Inner surface — **mucosal layer of cuboidal epithelium** continuous with middle ear mucosa

II.   Middle ear

   A.   Tympanic cavity

      1.   Space between tympanic membrane and labyrinth containing auditory ossicles and muscles, facial nerve, and openings to eustachian tube and mastoid air cells

      2.   Mucous membrane lining of tympanic cavity consists of simple cuboidal epithelium with occasional ciliated cells and goblet (mucin-containing) cells

   B.   **Mastoid air cells** — thin bony trabecular network of interconnecting air spaces stemming from tympanic cavity

   C.   **Eustachian tube**

      1.   Connects tympanic cavity (osseus end) and nasopharynx (cartilagenous end); lined by ciliated pseudostratified columnar (respiratory) epithelium

      2.   **Equalizes pressures on each side of tympanic membrane** by allowing air to enter tympanic cavity

      3.   In resting state, cartilagenous end is closed; during swallowing or yawning, contraction of palatal muscles pulls on cartilage causing tube to open

D. **Auditory ossicles** (middle ear bones) and muscles

    1.   Enhances transmission of sound energy from tympanic membrane to inner ear fluid (about 40 decibel amplification; dysfunction of ossicular chain produces conductive hearing loss of approximately this magnitude)

    2.   Three ossicles form lever mechanism

        a.   **Malleus** — manubrium (handle) of malleus merges inferiorly with central connective tissue layer of tympanic membrane

        b.   **Incus** — bone interposed between malleus and stapes

        c.   **Stapes** — footplate of stapes adjoins oval (vestibular) window (stapediovestibular joint) and is connected to cartilagenous rim of vestibular window by fibrous strands

    3.   Two muscles modify action of ossicles

        a.   **Tensor tympani muscle** — originates in bone near eustachian tube and connects to manubrium (handle) of malleus; contraction tenses tympanic membrane

        b.   **Stapedius muscle** — originates in bone near facial canal and attaches to stapes; contraction limits movement of ossicular chain as **protective reflex to loud sound**

E. **Facial nerve** (cranial nerve VII)

    1.   Passes along medial wall of tympanic cavity superior to oval window

    2.   Gives off nerve to stapedius muscle

    3.   Gives off **chorda tympani**

        a.   Efferent preganglionic parasympathetic fibers that stimulate secretion by submaxillary and sublingual glands

        b.   Afferent fibers for taste sensation from anterior two-thirds of tongue

III. **Inner ear**

A. **Labyrinth**

1. **Bony labyrinth** (otic capsule) — series of bony canals containing membranous labyrinth and surrounding perilymph

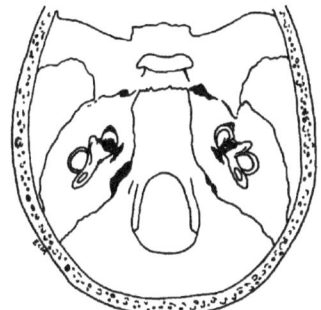

Relative positions of ear structures in petrous bone (viewed from interior of skull).

    a. Bone has **extremely hard consistency** due to presence within endochondral bone of calcified cartilagenous matrix and primitive bone that were not replaced during development and persist through adult life

    b. **Endosteum** — internal periosteal lining of bony labyrinth

2. **Perilymph**

    a. **Fluid interposed between endosteum of bony labyrinth and membranous labyrinth**

    b. **Continuous with subarachnoid cerebrospinal fluid** through cochlear aqueduct — canal connecting basal turn of cochlear scala tympani with subarachnoid space at jugular foramen adjacent to cranial nerve IX (glossopharyngeal nerve)

    c. Absorbed by cochlear spiral ligament

3. **Membranous labyrinth**

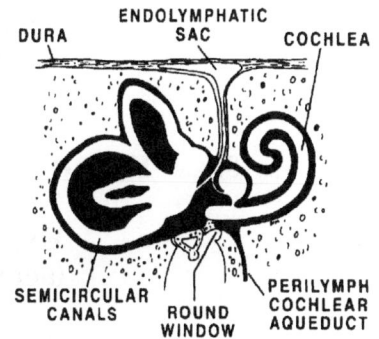

Labyrinthine structures.

    a. Complicated system of canals and ducts contained within bony labyrinth

    b. **Surrounded by perilymph**

    c. **Contains endolymph and delicate sensory organs** for hearing (organ of Corti) and balance (macula of utricle, macula of saccule, and ampulla of semicircular canal)

4. Endolymph

    a. **Fluid contained within membranous labyrinth**

    b. Electrolyte composition: high potassium and low sodium concentrations

    c. **Endolymphatic duct** — channel in bony vestibular aqueduct that connects utricle and saccule with endolymphatic sac

    d. **Endolymphatic sac** — terminal enlargement of endolymphatic duct within dural layers **adjacent to sigmoid sinus** that serves as **site for absorption of endolymph**

B. Cochlea

  1. **Osseus cochlea**

    a. **Spiral bony canal** containing **cochlear duct**

    b. Divided into two perilymph channels (**scala vestibuli** and **scala tympani**) that communicate at apex of cochlea

  2. **Oval window**

    a. **Extensible fibrous membrane attached to stapes footplate** and covering **opening at base of osseus cochlea (scala vestibuli)**

    b. Sound vibrations of tympanic membrane cause movement of auditory ossicles; back-and-forth **stapes movement initiates pressure waves of same frequency in perilymph of scala vestibuli**

  3. **Round window**

    a. **Extensible membrane covering opening in osseus cochlea (scala tympani)**

    b. Perilymph pressure waves initiated by oval window movement cause round window to bulge into middle ear or to retract

4. **Cochlear duct (scala media) — endolymph duct that contains organ of Corti**; separated from scala vestibuli by Reissner's membrane and from scala tympani by basilar membrane

    a. **Organ of Corti (hearing)**

        (1) Delicate structure **lying on basilar membrane** and protruding into endolymph of cochlear duct, consisting of **gelatinous tectorial membrane, sensory hair cells with apical stereocilia (sensory hairs)**, supporting cells, and nerve endings

        (2) **Traveling pressure wave in perilymph causes upward displacement of basilar membrane** with resultant distortion of hair cell stereocilia embedded in gelatinous tectorial membrane

        (3) Amplitude of basilar membrane displacement varies with sound intensity and frequency

             (a) **Basal turn has greater displacement with high frequency tones**, while apical turn has greater displacement with low frequency tones

             (b) Frequency responses of human basilar membrane range from 35 Hz to 20,000 Hz

        (4) Distortion of hair cell stereocilia produces potential changes transmitted to sensory (afferent) endings of cochlear nerve

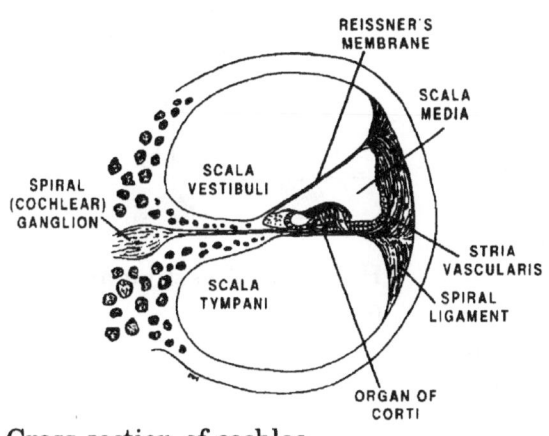

Cross section of cochlea.

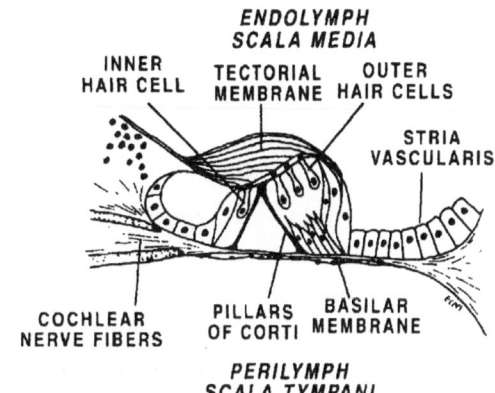

Organ of Corti.

C. **Vestibular system — saccule, utricle, and three semicircular ducts**

1. **Utricle and saccule (sensing position of head in space)**

    a. Macula

        (1) Sensory organ containing polarized sensory **hair cells** which have bundle of apical stereocilia and **single peripheral kinocilium**

        (2) Cilia are embedded in gelatinous **otolith membrane** composed of mucopoly-saccharide and weighted with calcium carbonate crystals

        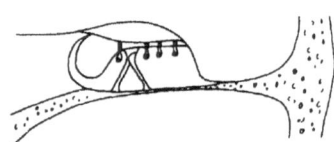

        (3) **Movement of head** causes **shifting of weighted otolith membrane** which causes distortion of hair cell cilia producing potential changes transmitted to sensory (afferent) endings of vestibular nerve

        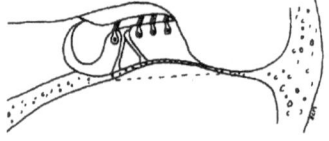

        Organ of Corti at rest (top) and with basilar membrane displacement by perilymph sound wave (bottom).

2. **Semicircular ducts (detecting rotatory head movement)**

    a. Three semicircular ducts (contained within bony semicircular canals)

        (1) Lateral (inclined 30 degrees downward from horizontal plane)

        (2) Posterior (oriented antero-posterior in sagittal plane)

        (3) Superior (oriented in coronal plane)

        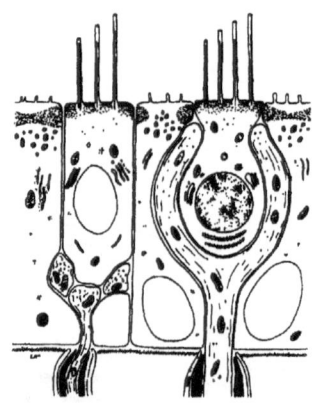

        Hair cells of vestibular organs.

    b. Sensory **hair cells** of crista of ampulla of each semicircular canal have **single apical kinocilium** located at periphery of bundle of stereocilia

    c.   Cilia are embedded in **cupula** composed of gelatinous mucopolysaccharide

    d.   **Rotatory movement of head** causes **fluid movement with displacement of cupula** resulting in distortion of hair cell cilia producing potential changes transmitted to sensory (afferent) endings of vestibular nerve

IV.   Physiologic studies

    A.   Vestibular function

        1.   **Nylen-Bárány maneuver** — patient seated on examining table is suddenly lowered to supine position with head thrust 45 degrees backward over end of table and turned 45 degrees to one side; development of vertigo or nystagmus (particularly asymmetric) suggests vestibular disease

        2.   **Caloric test**

            a.   **Unilateral vestibular stimulation** by instillation of cold or warm water into one external auditory meatus against tympanic membrane causing temperature changes in middle ear and in inner ear fluids

            b.   When patient lies flat (on back) with head flexed forward 30 degrees, **horizontal semicircular canal** is oriented vertically; inner ear temperature change causes endolymph to flow upward or downward stimulating sensory hair cells of ampullary crista with resultant nystagmus

                (1)   Warm water — slow movement of eyes away from stimulated ear with fast correction toward stimulated ear

                (2)   Cold water — slow movement of eyes toward stimulated ear and fast correction away from stimulated ear

        3.   **Electronystagmometry** — electrodes placed around eye record nystagmoid eye movements (either spontaneous or induced by rotational or caloric testing)

B.   Auditory function

1.   **Weber's test** — sound from vibrating tuning fork placed in **midline on patient's forehead** should be heard equally in both ears

a.   Unilateral sensorineural deafness — sound heard better in ear with normal acuity

b.   Unilateral conductive deafness — sound heard better in diseased ear

2.   **Rinne test** — vibrating tuning fork first is held in front of external auditory meatus (**air conduction**) and then stem of tuning fork is placed firmly against mastoid process (**bone conduction**)

a.   Conductive deafness — bone conduction louder than air conduction

b.   Normal individual or sensorineural deafness — air conduction louder than bone conduction

3.   **Audiometry**

a.   **Pure tone audiometry** — determines **loudness threshold** for hearing sounds (tones) of various frequencies

(1)   Conductive deafness — greater **low frequency** hearing loss

(2)   Sensorineural deafness — greater **high frequency** hearing loss

b.   **Speech discrimination** — inability to understand words despite normal (or nearly normal) pure tone audiometry; indicates **retrocochlear lesion (lesion of cochlear nerve or brain stem)**

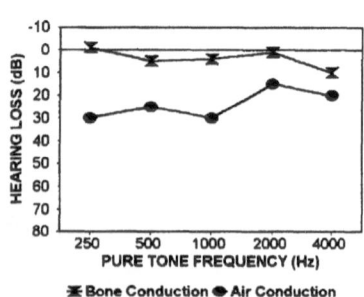

Audiogram of conductive hearing loss.

c.   **Alternate binaural loudness balance (ABLB)** — test of **recruitment** in which ear with hearing loss cannot detect low intensity sounds but hears loud sounds equally with good ear; indicates **cochlear disease**

d. **Short increment sensitivity index (SISI)** — determines ability to detect **small increases (increments) in loudness** while listening to simultaneous continuous tone; these sound increments are not heard with **retrocochlear lesion (lesion of cochlear nerve or brain stem)**

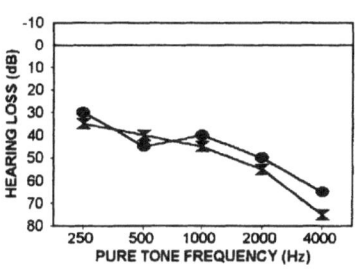

Audiogram of sensorineural hearing loss.

e. **Tone decay** — continuous tones become **progressively inaudible** with **cochlear nerve lesion**, while there is no change in loudness normally or with cochlear disease

f. **Impedance tympanometry** — measurement of ability of tympanic membrane to move in response to varying air pressures; excess movement suggests disruption of ossicular chain, while **reduced movement** suggests **otosclerosis** or **middle ear fluid collection**

g. **Stapedial reflex** — protective reflex in which **loud noise** in one ear normally causes stapedius muscles in both ears to contract; **reflex arc** consists of **cochlear nerve** (afferent limb of reflex), **brain stem interneurons**, and **facial nerve** (efferent limb of reflex); lesions of **cochlear nerve** often reduce or abolish reflex even when hearing is still normal

h. **Békésy audiograms** — measures perceived differences in loudness of continuous versus pulsed tones; type I tracing is normal, type II tracing suggests cochlear disease, while **type III** and **type IV** tracings indicate **retrocochlear lesion (lesion of cochlear nerve or brain stem)**

i. **Brain stem auditory evoked response (BAER)** — computerized averaging of scalp potentials produced by click sounds in each ear

    (1) Changes in **latency** or **amplitude** of recorded potentials suggests disease in peripheral or central auditory pathways

(2) Potential peaks — originate from specific sites in auditory pathway

(3) Latency between peaks (**interpeak latency**) is indicative of intactness of connecting tracts

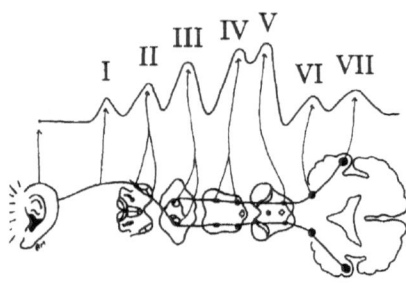

BAER potential peaks and responsible auditory pathway.

(4) Wave I — cochlear nerve

(5) Wave II — cochlear nuclei in upper medulla

(6) Wave III — trapezoid bodies and superior olivary nuclei bilaterally in lower pons

(7) Wave IV — lateral lemniscus and its nuclei in upper pons

(8) Wave V — midbrain inferior colliculi

(9) Wave VI — medial geniculate nuclei

(10) Wave VII — primary auditory cortex (Heschl's gyri)

V. Pathologic changes

A. **Otitis externa**

1. **Inflammation involving external auditory canal**

2. Usually related to growth of low virulence organisms in conditions of excessive moisture in canal (as in frequent swimming, hence "swimmer's ear")

B. **Exostoses** of external auditory canal — proximity of bone to skin surface makes bone susceptible to minor trauma with consequent **bony overgrowth** to form exostoses (for example: repeated cold water in external auditory canal of swimmers stimulates new bone formation)

C. **Basal cell carcinoma** or **squamous cell carcinoma** — particularly involves sun exposed portions of pinna

D. **Tympanic membrane abnormalities**

1. **Bullous myringitis — viral inflammation of tympanic membrane**
    with serum and blood collected between epidermal and connective tissue
    layers

2. **Perforation**

    a. Causes include trauma and infection

    b. Spontaneous healing common, but (depending on location and size
        of perforation) new replacement membrane may be thin and
        delicate or thick and fibrotic

E. **Otic barotrauma**

1. During **ascent to high altitude, increased pressure in tympanic
    cavity** (associated with outward bulging of tympanic membrane)
    passively forces open eustachian tube to allow air escape and
    equalization of pressures (associated with audible "click")

2. During **descent from high altitude, eustachian tube opening
    requires voluntary muscular activity** (swallowing or yawning); such
    opening may not occur because of anatomic blockade of eustachian tube
    or infection of middle ear or nasopharynx

    a. Low tympanic cavity pressure and high external air pressure
        results in inward bowing of tympanic membrane and edema of
        middle ear mucous membranes with transudation of fluid filling
        which fills tympanic cavity

    b. Rapid pressure changes can result in tympanic membrane
        rupture

    c. Associated with conductive hearing loss

F. **Otitis media**

1. **Acute suppurative otitis media** — bacterial inflammation involving
    middle ear; most common in children

    a. Inflammatory thickening of mucous membranes with impaired
        drainage through eustachian tube from

b. **Purulent otorrhea** — rupture of distended tympanic membrane from accumulation of purulent material in middle ear

c. **Mastoiditis** (involvement of mastoid air cells) produces bony destruction; potential complications include:

    (1) Injury to facial nerve with resultant facial palsy

    (2) Spread of infection to intracranial compartment

        (a) Brain abscess

        (b) Meningitis

        (c) Otitic hydrocephalus — lateral sinus thrombosis

2. **Serous otitis media**

a. Follows resolution of acute suppurative otitis media

b. **Persistent mucosal thickening prevents normal eustachian tube function** (air entry into tympanic cavity is blocked)

c. **Mucosal glandular metaplasia** results in accumulation of viscous seromucinous fluid

d. **Conductive hearing loss**

e. Treatment

    (1) Surgical placement of ventilating tube through tympanic membrane which allows mucosa to return to normal

    (2) With return of normal eustachian tube function, ventilating tube is spontaneously extruded, followed by healing of surgical tympanic membrane perforation

3. **Chronic suppurative otitis media**

a. Chronic infection associated with tympanic membrane perforation and purulent otorrhea

b. **Rarefying osteitis** results in **destruction of ossicles, mastoid bone, and otic capsule** with subsequent destruction of inner ear

    c.   **Cholesteatoma (keratoma)**

        (1)   **Accumulation of squamous debris in middle ear and mastoid air cells** from squamous epithelium that has invaded through tympanic membrane perforation

        (2)   Predisposes to bone destruction, damage to facial nerve, and spread of infection

        (3)   Ttreatment — surgical resection

G.   **Otosclerosis**

    1.   **Gradual replacement of normal bone** of bony labyrinth and stapes footplate by **lamellar new bone**

    2.   **Conductive hearing loss** of up to 40 decibels due to **progressive stapes immobility** from **fixation (fusion) of footplate to oval window annulus**

H.   **Basilar skull fracture** involving temporal bone of middle cranial fossa

    1.   Longitudinal fracture — fracture parallel to petrous pyramid produces damage to middle ear bones and tympanic membrane, bleeding from external auditory canal, and conductive hearing loss

    2.   **Transverse fracture — crosses petrous pyramid** with fracture through bony labyrinth and damage to membranous labyrinth with **sensorineural hearing loss, vestibular dysfunction,** and **cerebrospinal fluid otorrhea,** and possible damage to cranial nerve VIII (vestibulocochlear nerve) or cranial nerve VII (facial nerve)

        a.   **Cerebrospinal fluid otorrhea** — fractures through bony labyrinth allow perilymph (which is continuous with cerebrospinal fluid) to escape into middle ear (if tympanic membrane is intact, fluid drains through eustachian tube into nasopharynx)

        b.   **Infectious organisms** can **enter opening in bony labyrinth** and **ascend through perilymph to subarachnoid space producing meningitis**

   c. Fractures through bony labyrinth and endosteum heal poorly allowing prolonged leakage of cerebrospinal fluid and predisposing to repeated ascending infection

I. **Ménière's disease**

  1. Recurrent **sudden attacks of severe vertigo** (sensation of rotation of self or environment) lasting minutes to hours associated with **tinnitus** (ringing in ear), nausea and vomiting, and feeling of fullness or pressure in ear; **decrement of hearing** is also associated with each attack

  2. Frequency of attacks is variable and periods of remission can occur; vertiginous **attacks cease when sensorineural deafness is complete**

  3. Characterized by **massive dilation of endolymph spaces** ("**endolymphatic hydrops**") due to **abnormality of endolymph circulation** (normal flow from labyrinthine ducts into endolymphatic sac and subsequent absorption) resulting in distortion and rupture of membranous labyrinth and degeneration of hair cells

J. **Symptomatic endolymphatic hydrops**

  1. **Acute toxic labyrinthitis**

    a. **Vertigo and hearing loss** associated with meningitis or otitis media but **without bacterial invasion** of inner ear

    b. Accumulation of **proteinaceous debris in perilymph space** along with dilation of endolymph ducts

  2. **Acute suppurative labyrinthitis**

    a. **Bacterial invasion of labyrinth** from middle ear (through round or oval windows) or from subarachnoid space (through periotic duct in cochlear aqueduct)

    b. Inflammation in perilymph results in massive dilation of endolymph ducts and subsequent **destruction of membranous labyrinth**

K. **Viral labyrinthitis** — vertigo, tinnitus, or deafness developing in relation to viral illness due to viral damage to hearing or vestibular sensory organs

L. Drug toxicity — **ototoxic medications**

1. **Streptomycin** — relatively **selective destruction of hair cells of cristae** of semicircular canals

2. Dihydrostreptomycin, kanamycin, and gentamicin — less selective destruction of hair cells of organ of Corti (greatest toward basal area) with lesser destruction of hair cells of vestibular system

3. Quinine — loss of hair cells or complete destruction of organ of Corti with additional atrophy of striae vasculares

4. Salicylates — reversible dose-related tinnitus and hearing loss with no histopathologic change

M. **Presbycusis**

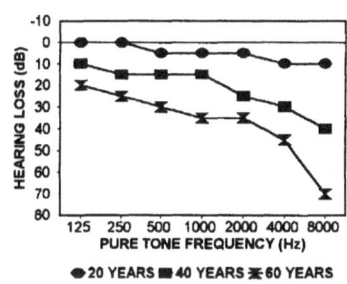

Audiogram of presbycusis.

1. Progressive age-related high frequency hearing loss

2. **Loss of hair cells and supporting cells of organ of Corti** progressing from basal end of cochlea and associated with concomitant loss of cochlear neurons

3. Progressive high frequency hearing loss with preserved speech discrimination

4. **Disequilibrium of aging** — vestibular system counterpart

N. **Acoustic neuroma (acoustic neurilemmoma or schwannoma)**

1. Insidious unilateral **hearing loss, tinnitus**, and **unsteadiness**; vertigo develops late in course of disease

2. Slow-growing, encapsulated benign posterior fossa tumor attached to cranial nerve VIII (vestibulocochlear nerve)

3. Tumor grows out of **internal auditory meatus** into **cerebellopontine angle** compressing adjacent cranial nerves (symptoms of facial numbness and weakness) and cerebellum (resulting in clumsiness or ataxia)

4. **Neurofibromatosis** — autosomal dominant **neurocutaneous syndrome** (**phakomatosis**) associated with acoustic neuromas

    a. Peripheral neurofibromatosis (neurofibromatosis type 1; NF-1; classical neurofibromatosis; von Recklinghausen's syndrome) — multiple **café-au-lait spots**, **Lisch nodules** (iris hamartomas appearing as small yellow or brown elevations), cutaneous neurofibromas, spinal and cranial nerve root neurofibromas or neurilemmomas, and skeletal anomalies; gene localized to chromosome 17

    b. Central neurofibromatosis (neurofibromatosis type 2; NF-2) — **bilateral acoustic neuromas** with few café-au-lait spots; gene localized to chromosome 22

5. Diagnosis made with audiometry, brain stem auditory evoked potentials, and radiologic imaging studies; complete surgical resection is necessary

SUGGESTED ADDITIONAL READING

Goodhill V: *Ear Diseases, Deafness, and Dizziness.* Hagerstown, Md., Harper & Row, 1979.

Schuknecht HF: *Pathology of the Ear.* Cambridge, Mass., Harvard University Press, 1974.

Schuknecht HF, Gulya AJ: *Anatomy of the Temporal Bone with Surgical Implications.* Philadelphia, Lea & Febiger, 1986.

SELF-ASSESSMENT EXAMINATION #1

1. Abnormality of peroxisomal function is found in:
   A. Morquio syndrome
   B. Alexander's disease
   C. Metachromatic leukodystrophy
   D. Adrenoleukodystrophy
   E. Tay-Sachs disease

2. Growth of squamous epithelium that has migrated into the middle ear through a perforation in the tympanic membrane produces:
   A. Cholesteatoma
   B. Bullous myringitis
   C. Otitic barotrauma
   D. Serous otitis media
   E. Otosclerosis

3. Absence of the muscle membrane protein dystrophin characterizes which disorder:
   A. Myasthenia gravis
   B. Infantile spinal muscular atrophy
   C. Myotonic muscular dystrophy
   D. Duchenne muscular dystrophy
   E. Myoclonus epilepsy with ragged-red fibers (MERRF)

4. Which of the following features are associated with anencephaly:
   A. Oligohydramnios
   B. Absence of cerebellar cortex
   C. Hypoplastic adrenal glands
   D. Absence of cranial nerves III, IV, VI
   E. Hypoplastic anterior lobe of pituitary gland

5. Rosenthal fibers are characteristically found in which tumor:
   A. Oligodendroglioma
   B. Colloid cyst
   C. Medulloblastoma
   D. Meningioma
   E. Juvenile pilocytic astrocytoma

6. Neuritic (senile) plaques are considered specific for which neurodegenerative disease:
   A. Pick's disease
   B. Alzheimer's disease
   C. Creutzfeldt-Jakob disease
   D. Progressive supranuclear palsy
   E. Ataxia-telangiectasia

7. Relatively localized frontotemporal atrophy ("lobar atrophy") characterizes which neurodegenerative disorders:
   A. Parkinson's disease
   B. Alzheimer's disease
   C. Huntington's disease
   D. Pick's disease
   E. Progressive supranuclear palsy

8. A tumor characteristic of von Hippel-Lindau disease is:
   A. Optic glioma
   B. Pontine glioma
   C. Subependymal giant cell astrocytoma
   D. Cerebellar hemangioblastoma
   E. Acoustic neuroma

9. Characteristics of pineal tumors include all the following EXCEPT:
   A. Associated with Parinaud's (dorsal midbrain) syndrome
   B. Produce distant symptoms from subarachnoid dissemination
   C. Can secrete α-fetoprotein or human chorionic gonadotrophin
   D. Produces symptoms from local pressure on cranial nerve VIII
   E. Associated with obstructive hydrocephalus

10. Pigmentary retinopathy with progressive dementia characteristics:
    A. Ceroid lipofuscinosis
    B. Gaucher's disease
    C. Fabry's disease
    D. Phenylketonuria
    E. Zellweger's syndrome

11. The report from the clinical neurophysiology laboratory on a 25 year old woman complaining of muscle weakness shows progressive decrement in amplitude of muscle action potential recorded during 3 Hz repetitive nerve stimulation. The most likely diagnosis is:
    A. Myotonic dystrophy
    B. Duchenne muscular dystrophy
    C. Charcot-Marie-Tooth disease
    D. Myasthenia gravis
    E. Guillain-Barré syndrome

12. The cavernous sinus contains all the following structures EXCEPT:
    A. Oculomotor nerve
    B. Trochlear nerve
    C. Facial nerve
    D. Trigeminal nerve
    E. Abducens nerve

13. All the following are characteristic features of the Arnold-Chiari malformation EXCEPT:
    A. Elongated unrolled cerebellar vermis
    B. Kinked medulla
    C. Beaked collicular plate
    D. Pachygyria
    E. Aqueductal stenosis

14. Presbycusis is:
    A. Age-related high frequency hearing loss
    B. Sudden attacks of severe vertigo
    C. Conductive hearing loss
    D. Chronic otorrhea
    E. Tinnitus

15. Epidural hematoma results from damage to which vascular structure:
    A. Middle meningeal artery
    B. Superior sagittal sinus
    C. Bridging cerebral cortical veins
    D. Middle cerebral artery
    E. Internal cerebral vein

16. A 3 month old infant dies in its crib. The child had been born prematurely at 31 weeks gestational age. At autopsy the brain shows yellow discoloration of the basal ganglia, thalamus, inferior olivary nuclei, and cerebellar dentate nuclei. The most likely explanation of this finding is:
    A. Tay-Sach's disease
    B. Kernicterus
    C. Alexander's disease
    D. Krabbe's disease
    E. Adrenoleukodystrophy

17. The most common cause of blindness worldwide is due to infection by:
    A. Herpes simplex
    B. Syphilis
    C. Trachoma
    D. *Neisseria gonorrhea*
    E. *Haemophilus influenzae*

18. Chemical meningitis from oxalate crystal deposition in the brain occurs following poisoning with which substance:
    A. Methanol
    B. Ethylene glycol
    C. Arsenic
    D. Organic mercury
    E. Ethanol

19. Sulfatide accumulation occurs in:
    A. Canavan's disease (spongy degeneration)
    B. Metachromatic leukodystrophy
    C. Parkinson's disease
    D. Niemann-Pick disease
    E. Pick's disease

20. Slow growing benign spinal tumors are associated with the sudden onset of spastic paraplegia due to:
    A. Spinal cord ischemia
    B. Distant tumor metastases
    C. Anterior horn cell loss
    D. Spinal nerve root compression
    E. Demyelination of dorsal columns

21. A 10 year old child with severe cystic fibrosis who develops peripheral neuropathy and ataxia is most likely suffering from:
    A. Vitamin B_{12} deficiency
    B. α-Tocopherol deficiency
    C. Lead poisoning
    D. Central pontine myelinolysis
    E. Hepatic encephalopathy

22. Which of the following is most likely to produce a chronic basilar meningitis:
    A. *Mycobacterium tuberculosis*
    B. *Neisseria meningitidis*
    C. *Haemophilus influenzae*
    D. *Escherichia coli*
    E. *Listeria monocytogenes*

23. A 64 year old woman presents with complaints of tingling in her legs and difficulty walking. Examination shows spasticity and loss of position sensation in the legs. Her blood studies reveal a macrocytic anemia. The most likely cause of her problem is:
    A. Vitamin B_{12} deficiency
    B. Hypercholesterolemia
    C. Congophilic angiopathy
    D. Tissue copper deposition
    E. Lead poisoning

24. Gaucher's disease is characterized by deficiency of which enzyme:
    A. Glucocerebroside-ß-glucosidase
    B. Sphingomyelinase
    C. Hexosaminidase A
    D. Arylsulfatase A
    E. Galactocerebroside-ß-galactosidase

25. The tumor of the clivus arising from remnants of embryonic notochord is:
    A. Neurofibroma
    B. Meningioma
    C. Medulloblastoma
    D. Hemangioblastoma
    E. Chordoma

26. The fastest conducting peripheral nerve axons are:
    A. Efferents to intrafusal muscle fibers
    B. Preganglionic autonomic efferents
    C. Afferents from pain receptors
    D. Afferents from Ruffini corpuscles
    E. Afferents from Golgi tendon organs

27. Tumor enlargement of the pineal gland would result in pressure on which structures:
   A. Pituitary gland
   B. Superior colliculi
   C. Facial colliculi
   D. Substantia innominata
   E. Genu of corpus callosum

28. A 30 year old man has the relatively abrupt onset of seizures and confusion. Lumbar puncture shows blood-tinged cerebrospinal fluid and an MRI scan suggests hemorrhagic necrosis of the left anterior temporal lobe with cerebral edema and early transtentorial uncal herniation. The most likely diagnosis is:
   A. Rabies
   B. Poliomyelitis
   C. Subacute sclerosing panencephalitis
   D. Herpes simplex encephalitis
   E. Progressive multifocal leukoencephalopathy

29. Bilateral carotid artery occlusion in utero leads to:
   A. Chiari type II malformation
   B. Hydranencephaly
   C. Werdnig-Hoffman disease
   D. Lissencephaly syndrome
   E. Dandy-Walker malformation

30. Tay-Sachs disease is characterized by deficiency of which enzyme:
   A. Glucocerebroside-ß-glucosidase
   B. Sphingomyelinase
   C. Hexosaminidase A
   D. Arylsulfatase A
   E. Galactocerebroside-ß-galactosidase

31. Which of the following is the major inhibitor of prolactin release:
   A. Somatostatin
   B. Vasoactive intestinal peptide
   C. Serotonin
   D. Norepinephrine
   E. Dopamine

32. Massive cerebral edema in children with pica is due to poisoning with which substance:
   A. Lead
   B. Mercury
   C. Thiamine
   D. Ferrous sulfate
   E. Ethanol

33. Relatively selective loss of GABAergic neurons of the striatum characterizes which neurodegenerative disorder
   A. Pick's disease
   B. Alzheimer's disease
   C. Huntington's disease
   D. Binswanger's disease
   E. Friedreich's ataxia

34. Compromise of the vasa nervorum with nerve infarction of numerous peripheral nerves in polyarteritis nodosum produces the clinical syndrome of:
   A. Autonomic neuropathy
   B. Motor neuropathy
   C. Mononeuropathy multiplex
   D. Sensory neuropathy
   E. Shy-Drager syndrome

35. Excess urinary excretion of heparan sulfate occurs in:
   A. Phenylketonuria
   B. Homocystinuria
   C. Hurler syndrome
   D. Down's syndrome
   E. Metachromatic leukodystrophy

36. The non-infectious granulomatous inflammation around lipid droplets of sebaceous glands in the eyelids is known as:
   A. Entropion
   B. Chalazion
   C. Orbital pseudotumor
   D. Keratomalacia
   E. Kayser-Fleischer ring

37. The vascular abnormality frequently associated with Alzheimer's disease is:
    A. Polyarteritis nodosa
    B. Temporal arteritis
    C. Fibromuscular dysplasia
    D. Moyamoya disease
    E. Cerebral amyloid angiopathy

38. A 24 year old woman has for many years had intermittent headaches, particularly after lying in bed. CT scan indicates lateral ventricular enlargement and a third ventricular tumor. The most likely diagnosis is:
    A. Colloid cyst
    B. Medulloblastoma
    C. Hemangioblastoma
    D. Schwannoma
    E. Neurofibroma

39. A 60 year old man complains of frequent sudden attacks of severe vertigo and tinnitus. Between attacks he has noted some loss of hearing. Which of the following is the most likely diagnosis:
    A. Otitic barotrauma
    B. Acoustic schwannoma
    C. Tuberous sclerosis
    D. von Hippel-Lindau disease
    E. Meniere's disease

40. Demyelination is characterized by the formation of:
    A. Sudanophilic fats
    B. Congophilic amyloid
    C. Neurofibrillary tangles
    D. Lewy bodies
    E. Paracrystalline inclusions

41. Lewy bodies are found in which neurodegenerative disorder:
    A. Parkinson's disease
    B. Pick's disease
    C. Friedreich's ataxia
    D. Huntington's disease
    E. Ataxia-telangiectasia

42. Too rapid correction of hyponatremia results in:
    A. Alzheimer type II astrocytosis
    B. Distal axonal degeneration
    C. Pigmentary retinopathy
    D. Central pontine myelinolysis
    E. Pallidal necrosis

43. The uveal tract consists of all the following EXCEPT:
    A. Iris
    B. Choroid
    C. Ciliary body
    D. Pupillary dilator muscle
    E. Cloquet's canal

44. Subdural hematoma results from damage to which vascular structure:
    A. Middle meningeal artery
    B. Bridging cerebral cortical veins
    C. Middle cerebral artery
    D. Cavernous sinus
    E. Internal cerebral vein

45. The specialized sensory receptors that are encapsulated nerve endings of large myelinated fibers involved in fine discrimination touch sensation on palmar skin are:
    A. Meissner corpuscles
    B. Pacinian corpuscles
    C. Golgi-Mazzoni endings
    D. Krause end bulbs
    E. Golgi tendon organs

46. The early memory disturbance in Alzheimer's disease relates to loss of neurons in which structure:
    A. Clarke's column
    B. Nucleus basalis of Meynert
    C. Locus ceruleus
    D. Substantia nigra
    E. Subthalamic nucleus

47. Myelin in the central nervous system is produced by which cell:
    A. Meningothelial cells
    B. Astrocytes
    C. Schwann cells
    D. Oligodendrocytes
    E. Perineurial fibroblasts

48. Anterograde amnesia and confabulation are characteristic clinical features of:
    A. Progressive supranuclear palsy
    B. Hepatic encephalopathy
    C. Vitamin B_{12} deficiency
    D. Pick's disease
    E. Korsakoff's psychosis

49. What are the specialized sensory receptors that consist of concentric layers of cells and fluid surrounding large myelinated nerve fibers in subcutaneous tissue, fascia, and mesentery?
    A. Meissner corpuscles
    B. Pacinian corpuscles
    C. Golgi-Mazzoni endings
    D. Krause end bulbs
    E. Golgi tendon organs

50. Possible late delayed complications of head injury include all the following EXCEPT:
    A. Epilepsy
    B. Hydrocephalus
    C. Personality changes
    D. Duret hemorrhage
    E. Neurofibrillary tangles

SELF-ASSESSMENT EXAMINATION #2

1. Band keratopathy is due to:
   A. Calcium deposition in Bowman's membrane
   B. Deficient tear production
   C. Copper sulfide deposition in Descemet's membrane
   D. Chlamydia infection
   E. Impaired aqueous humor formation

2. A floppy dysmorphic neonate with high narrow forehead, flat nose, corneal opacities, congenital heart disease, jaundice, splenomegaly, renal cystic dysplasia, and genital anomalies most probably has:
   A. Zellweger's syndrome
   B. Gaucher's disease
   C. Leigh's disease
   D. Canavan's disease
   E. Alexander's disease

3. Brain biopsy in a 30 year old patient with AIDS and mental confusion shows granular destruction of the white matter with slight chronic inflammatory cell infiltration, enlarged oligodendroglial nuclei, and bizarre giant astrocytic nuclei. The most likely diagnosis is:
   A. Neurosyphilis
   B. Subacute sclerosing panencephalitis
   C. *Mycobacterium tuberculosis* infection
   D. Cerebral lymphoma
   E. Progressive multifocal leukoencephalopathy

4. Neurofibrillary tangles in basal ganglia, thalamus, and brain stem but none in cerebral cortex is characteristic of which neurodegenerative disease:
   A. Pick's disease
   B. Creutzfeldt-Jakob disease
   C. Progressive supranuclear palsy
   D. Huntington's disease
   E. Parkinson's disease

5. Devic's disease (neuromyelitis optica) is characterized by acute demyelinating plaques involving:
   A. Spinal cord
   B. Thalamus
   C. Fornix
   D. Anterior commissure
   E. Cerebellum

6. An infant with macrocephaly, progressive loss of developmental milestones, seizures, and brain biopsy showing innumerable Rosenthal fibers has:
   A. Leigh's disease
   B. Canavan's disease
   C. Alexander's disease
   D. Zellweger's syndrome
   E. Gaucher's disease

7. Pituitary corticotroph cells contain:
   A. Pro-opiomelanocortin
   B. Somatostatin
   C. Dopamine
   D. Cortisol
   E. Vasopressin

8.. The retinal examination of a 44 year old woman shows numerous tiny capillary aneurysms and hemorrhages along with neovascularization along the retinal surface. The most likely diagnosis is:
   A. Diabetic retinopathy
   B. Hypertensive retinopathy
   C. Optic neuritis
   D. Retinitis pigmentosa
   E. Retinoblastoma

9. A 32 year old woman has a long-standing temporal lobe seizure disorder that has recently become refractory to anti-epileptic drugs. CT scan shows a calcified temporal lobe mass. The most likely diagnosis is
   A. Oligodendroglioma
   B. Craniopharyngioma
   C. Colloid cyst
   D. Medulloblastoma
   E. Hemangioblastoma

10. A 55 year old man complains of transient excruciating pains in his legs. On exam, he has greatly reduced proprioception and pain sensibility in his legs. Radiographs of his knees reveal marked deformity and joint destruction. The most likely diagnosis is:
   A. Tabes dorsalis
   B. Poliomyelitis
   C. Progressive multifocal leukoencephalopathy
   D. Myotonic dystrophy
   E. Porphyria

11. A tumor characteristic of tuberous sclerosis is:
   A. Optic glioma
   B. Pontine glioma
   C. Subependymal giant cell astrocytoma
   D. Cerebellar hemangioblastoma
   E. Acoustic neuroma

12. Myelin in the peripheral nervous system is produced by which cells:
   A. Meningothelial cells
   B. Astrocytes
   C. Schwann cells
   D. Oligodendrocytes
   E. Perineurial fibroblasts

13. A 15 year old boy presents at the hospital with meningitis and a petechial skin rash. The most likely causative organism is:
   A. *Mycobacterium tuberculosis*
   B. *Neisseria meningitidis*
   C. *Haemophilus influenzae*
   D. *Escherichia coli*
   E. *Listeria monocytogenes*

14. Material moved by anterograde fast axoplasmic transport in peripheral nerve axons includes:
   A. Mitochondria
   B. Peptides
   C. Lysosomes
   D. Synaptic vesicles
   E. Golgi apparatus

15. Degeneration of spinal cord posterior columns and dorsal and ventral spinocerebellar tracts characterizes which disorder:
   A. Huntington's disease
   B. Friedreich's ataxia
   C. Parkinson's disease
   D. Olivopontocerebellar atrophy
   E. Shy-Drager syndrome

16. Caudate atrophy characterizes which neurodegenerative disorder:
   A. Parkinson's disease
   B. Pick's disease
   C. Alzheimer's disease
   D. Binswanger's disease
   E. Huntington's disease

17. A 60 year old man complains of a painful lump between his third and fourth toes. He has had several injuries and fractures involving that foot in the past. The most likely diagnosis is:
    A. Morton's neuroma
    B. Guillain-Barré syndrome
    C. Amyotrophic lateral sclerosis
    D. Porphyria
    E. Charcot-Marie-Tooth disease

18. A 2 year old child has right ear serous otitis media. The most likely symptom associated with this condition would be:
    A. Sensorineural hearing loss
    B. Conductive hearing loss
    C. Facial nerve palsy
    D. Cerebrospinal fluid otorrhea
    E. Endolymphatic hydrops

19. Polycythemia from tumor elaboration of an erythropoietin-like factor is typical of:
    A. Neurofibroma
    B. Cerebellar hemangioblastoma
    C. Subependymal giant cell astrocytoma
    D. Metastatic melanoma
    E. Ependymoma

20. The multisystem disorder characterized by distal muscle weakness, cataracts, frontal baldness, and cardiac arrhythmias is:
    A. Amyotrophic lateral sclerosis
    B. Subacute sclerosing panencephalitis
    C. Myotonic dystrophy
    D. Werdnig-Hoffman disease
    E. Charcot-Marie-Tooth disease

21. Battle's sign is the result of fracture of
    A. Petrous bone
    B. First cervical vertebrae
    C. Cribriform plate
    D. Zygomatic arch
    E. Floor of sella turcica

22. The common location for surface contrecoup contusions is:
    A. Cervical spinal cord
    B. Cerebellar tonsils
    C. Undersurface of frontal and temporal lobes
    D. Basis pons
    E. Posterior thalamus

23. MPTP toxicity is most similar to which neurodegenerative disorder:
    A. Parkinson's disease
    B. Friedreich's ataxia
    C. Ataxia-telangiectasia
    D. Pick's disease
    E. Wilson's disease

24. One neuropathologic pattern of neurologic damage in survivors of carbon monoxide poisoning is:
    A. Hemorrhagic necrosis of mamillary bodies
    B. Neuronal loss and gliosis in dorsomedial thalamus
    C. Necrosis of globus pallidus
    D. Shrinkage of anterior lobe cerebellar folia
    E. Alzheimer type II astrocytosis in gray matter

25. Wernicke-Korsakoff syndrome is due to deficiency of:
    A. Vitamin B_{12}
    B. Thiamine
    C. L-Carnitine
    D. Vitamin E
    E. Pyridoxine

26. The retinal pigmentary epithelium is separated from the choriocapillaris by:
    A. Descemet's membrane
    B. Bowman's membrane
    C. Tenon's capsule
    D. Bruch's membrane
    E. Internal limiting membrane of retina

27. Focal demyelination (with relative preservation of axons) in the anterior midline corpus callosum is found in:
    A. Vitamin B_{12} deficiency
    B. Thiamine deficiency
    C. Central pontine myelinolysis
    D. Marchiafava-Bignami disease
    E. Lead poisoning

28. Fabry's disease is characterized by:
    A. Burning pain in hands and feet
    B. Cherry red macular spot
    C. Hepatosplenomegaly
    D. Autofluorescent storage material
    E. Dwarfism and grotesque facial features

29. The only mucopolysaccharidosis with an X-linked (sex-linked) inheritance pattern is:
    A. Tay-Sachs disease
    B. Zellweger's syndrome
    C. Lesch-Nyhan syndrome
    D. Hunter syndrome
    E. Canavan's disease

30. Wallerian degeneration results from:
    A. Extrajunctional acetylcholine receptor formation
    B. Axonal disruption
    C. Neuromuscular junction blockade
    D. Immunologic attack against CNS myelin
    E. Viral infection of Schwann cells

31. Shadow plaques are:
    A. Thalamic plaques
    B. Plaques with axonal destruction
    C. Area of remyelination
    D. Peripheral nerve demyelination
    E. Found in acute necrotizing hemorrhagic encephalomyelitis

32. A 53 year old man experiences an 18 month course of progressive muscle weakness and wasting with fasciculations in nearly all muscles of the arms and legs, hyperactive tendon reflexes, and no sensory abnormality. Muscle biopsy of the quadriceps muscle would reveal:
    A. Selective type I fiber atrophy
    B. Muscle fiber necrosis and myophagocytosis
    C. Marked endomysial fibrosis
    D. Mitochondrial paracrystalline inclusions
    E. Fiber type grouping and grouped atrophy

33. Prenatal detection of anencephaly is possible by finding:
    A. Elevated amniotic fluid α-fetoprotein levels
    B. Low maternal HCG levels
    C. Low amniotic fluid acetylcholinesterase levels
    D. Low amniotic fluid hexosaminidase A levels
    E. Elevated maternal serum very long chain fatty acid levels

34. Defective pigment formation with fair skin, blond hair, and blue eyes is observed in:
    A. Lesch-Nyhan syndrome
    B. Wilson's disease
    C. Phenylketonuria
    D. Acid maltase deficiency
    E. Gaucher's disease

35. A 3 year old child presents with a 2 week history of vomiting, ataxia, and lethargy. MRI scans showed marked lateral and third ventricular enlargement with a mass filling the fourth ventricle. The most likely diagnosis is:
   A. Medulloblastoma
   B. Schwannoma
   C. Neurofibroma
   D. Hemangioblastoma
   E. Meningioma

36. Down's syndrome patients reaching age 30 years have histologic evidence in the brain of:
   A. Huntington's disease
   B. Wilson's disease
   C. Alzheimer's disease
   D. Progressive multifocal leukoencephalopathy
   E. Chiari type II malformation

37. The spinal condition of two hemicords separated by a bony spur is termed:
   A. Syringomyelia
   B. Hydromyelia
   C. Amyotrophic lateral sclerosis
   D. Meningocele
   E. Diastematomyelia

38. A generally healthy 52 year old man has always prided himself on having the "perfect tan", and for the past 20 years, he has been the proprietor of a tanning salon. He has recently developed visual blurring. Which of the following is the most likely explanation for his visual problem:
   A. Nuclear cataract
   B. Cortical cataract
   C. Retinal detachment
   D. Retinitis pigmentosa
   E. Optic neuritis

39. Melatonin
   A. Inhibits gonadal development and sexual maturity
   B. Stimulates release of gonadotrophins
   C. Is synthesized from dopamine
   D. Is present in pineal astrocytes
   E. Levels increase with light exposure

40. Characteristics of vasopressin include all of the following EXCEPT:
   A. Synthesized by hypothalamic neurons
   B. Deficiency leads to diabetes insipidus
   C. Inappropriate excessive secretion leads to hyponatremia
   D. Stimulates uterine contractions
   E. Ectopic production occurs with lung cancer

41. White matter cavitation with perivascular clusters of large macrophages filled with PAS-positive material characteristics:
   A. Krabbe's disease (globoid cell leukodystrophy)
   B. Tay-Sachs disease
   C. Alexander's disease
   D. Canavan's disease (spongy degeneration)
   E. Wilson's disease

42. What are the specialized sensory receptors that consist of free nerve endings ramifying around collagen bundles in tendons?
   A. Meissner corpuscles
   B. Pacinian corpuscles
   C. Golgi-Mazzoni endings
   D. Krause end bulbs
   E. Golgi tendon organs

43. Adults poisoned with which substance develop a motor neuropathy often presenting as wrist drop or foot drop:
    A. Arsenic
    B. Lead
    C. Methanol
    D. Thallium
    E. Carbon monoxide

44. Blindness from optic nerve and retinal damage is particularly prominent in survivors of poisoning with which substance:
    A. Arsenic
    B. Thallium
    C. Ethylene glycol
    D. Carbon monoxide
    E. Methanol

45. Twitchings of single denervated muscle fibers are termed:
    A. Myokymia
    B. Fasciculations
    C. Myotonia
    D. Fibrillations
    E. Dystonia

46. The congenital central nervous system malformation often associated with trisomy 13 (Patau's syndrome) is:
    A. Holoprosencephaly
    B. Hydranencephaly
    C. Dandy-Walker malformation
    D. Myelomeningocele
    E. Occipital encephalocele

47. An initially normal infant who during the first year of life develops hyperacusis, loss of developmental milestones, opisthotonus, and macular cherry-red spot probably has:
    A. Tay-Sachs disease
    B. Metachromatic leukodystrophy
    C. Adrenoleukodystrophy
    D. Fabry's disease
    E. Ceroid lipofuscinosis

48. A 32 year old woman complains of unilateral hearing loss, tinnitus, and unsteadiness. Exam shows multiple café-au-lait spots. The most likely diagnosis for her complaints are:
    A. Otitic barotrauma
    B. Acoustic schwannoma
    C. Tuberous sclerosis
    D. von Hippel-Lindau disease
    E. Ménière's disease

49. A 65 year old woman complains of recent onset of headache, malaise, and muscle and joint aches. Her erythrocyte sedimentation rate is 83 mm/hour. The most likely diagnosis is:
    A. Temporal arteritis
    B. Moyamoya disease
    C. Status marmoratus
    D. Myotonic dystrophy
    E. Amyotrophic lateral sclerosis

50. Relatively selective loss of dopaminergic fibers characterizes which neurodegenerative disorder:
    A. Parkinson's disease
    B. Pick's disease
    C. Amyotrophic lateral sclerosis
    D. Huntington's disease
    E. Wilson's disease

51. Metachromatic leukodystrophy is characterized by deficiency of which enzyme:
    A. Glucocerebroside-ß-glucosidase
    B. Sphingomyelinase
    C. Hexosaminidase A
    D. Arylsulfatase A
    E. Galactocerebroside-ß-galactosidase

Answers to Self-Assessment Examination #1

1. D	11. D	21. B	31. E	41. A
2. A	12. C	22. A	32. A	42. D
3. D	13. D	23. A	33. C	43. E
4. C	14. A	24. A	34. C	44. B
5. E	15. A	25. E	35. C	45. A
6. B	16. B	26. E	36. B	46. B
7. D	17. C	27. B	37. E	47. D
8. D	18. B	28. D	38. A	48. E
9. D	19. B	29. B	39. E	49. B
10. A	20. A	30. C	40. A	50. D

Answers to Self-Assessment Examination #2

1. A	11. C	21. A	31. C	41. A
2. A	12. C	22. C	32. E	42. E
3. E	13. B	23. A	33. A	43. B
4. C	14. B	24. C	34. C	44. E
5. A	15. B	25. B	35. A	45. D
6. C	16. E	26. D	36. C	46. A
7. A	17. A	27. D	37. E	47. A
8. A	18. B	28. A	38. B	48. B
9. A	19. B	29. D	39. A	49. A
10. A	20. C	30. B	40. D	50. A
				51. D